ments in eco-efficiency. Second, it showed this challenge to be far beyond the range of improvement possible through end-of-pipe technologies and even most 'environmental' technologies, which represent process-integrated improvements to the eco-efficiency of current processes and products. The study reported that, typically, these changes to established technologies could be expected to deliver improvements in eco-efficiency of no more than a factor of 2–3. Third, and most importantly, the corollary drawn was that, to have any chance of meeting the technological challenge of sustainability, it may be inadvisable to follow the traditional route of studying present products and processes as a way of identifying areas for improvement. Rather, it may be better to begin by accepting that sustainable technologies will most often consist of path-breaking approaches to meeting needs that are radically different from the solutions we have in place today.

The feasibility study thus introduced a number of programme-defining concepts. For the first time, it defined 'sustainable technologies' and differentiated these from 'end-of-pipe' approaches and 'environmental technologies'. Sustainable technologies would be ones capable of meeting 'needs' using only a fraction—less than a tenth and maybe only a fiftieth—of the 'eco-capacity' used by today's technologies. Eco-capacity was, itself, introduced as a concept to describe the constraints on the permissible level of resource consumption, ecosystem disruption and pollutant emission that would be consistent with maintaining a stock of environmental capital and a stream of environmental benefits for the use of future generations. The concept of 'need' was adopted from the Brundtland Commission's definition of sustainability (WCED 1987) and its status in innovation practice elevated so that it would be the starting point for innovation processes. This was seen necessary as a means of focusing long-term innovation on strategic issues and of combating the tendency for incrementalism. The factor concept was introduced as a means of operationalising a quantitative goal for reduction in resource use and improvement in eco-efficiency—with progress judged against the benchmark of the demand on eco-capacity made by today's solutions to meeting needs. The factor approach and benchmarking also make it possible to specify a time-path for eco-efficiency improvement, which can be translated into rates of change and used to monitor progress.

Above all, in order to facilitate a process in which the present situation plays little or no role in long-term innovation, a 'backcasting' approach—first introduced into the sustainability arena for end-use-oriented energy systems planning (Goldemberg *et al.* 1985)—was introduced. Backcasting begins with an attempt to envision an acceptable future system state, which takes into account the status of as many important defining constraints and criteria as possible, including the requirement to meet 'needs'. This system state is then used as a reference: for tracing pathways back to the present, for placing milestones along those pathways and for identifying short-term challenges and obstacles that will have to be overcome en route. Progress will depend not only on meeting the technological challenges, but also on co-evolutionary developments in policies, markets, attitudes and behaviours. R&D and innovation efforts have to be directed to all of these challenges. Backcasting thus provides a way of connecting the future and the present. It provides a means of translating a long-term vision of a sustainable future into near-term actions consistent both with achieving that future and dealing with the realities of the present situation. It also provides a basis for a co-evolutionary approach to innovation in which the various elements of the developmental system are viewed holistically and dynamically in terms of interrelationships and feedbacks.

ECO-RESTRUCTURING AND TECHNOLOGICAL INNOVATION

Other considerations also contributed to the decision to try to influence innovation practices. Long-term sustainability depends on eco-restructuring to bring the so-called 'metabolism' of our societies and economies—the amount and structure of resource use and waste production—within the boundary conditions described by critical eco-capacities and by human capacities to cope with environmental change or to accept it.[3] As a process, eco-restructuring implies achieving wide-ranging changes in our societies and economies including, especially, a restructuring of production and consumption patterns both in amount and type. This is an inevitable corollary of the important Ehrlich Identity (Ehrlich and Holdren 1971; Holdren and Ehrlich 1974; Ehrlich *et al.* 1977; Ehrlich and Ehrlich 1990).[4] Eco-restructuring also implies cultural and social restructuring, especially of values, motivations and institutions that underlie the criteria used when making produc-

---

3   It is therefore directly relevant to two of the three central sustainability concerns (for securing a continuing stream of environmental benefits and a continuing stream of economic benefits) and indirectly relevant to the third (for securing a more equitable distribution of entitlement to these benefits across and within generations).

4   Ehrlich's first expression for the important relationship between environmental impact and human activity was $I = P.F$ where I is environmental impact, P is population and F is impact per capita (Ehrlich and Holdren 1971). This equation was later expanded to $I = P.C.T$ where C is consumption per capita and T is the impact per unit of consumption, referred to elsewhere as the environmental impact coefficient (Holdren and Ehrlich 1974). The equation has been explored in works by its author, such as *Ecoscience* (Ehrlich *et al.* 1977) and *The Population Explosion* (Ehrlich and Ehrlich 1990), and has been widely used since to gain insight into the scale of the technological challenge posed by sustainability under different assumptions about eco-capacity, GNP growth/distribution and population growth/distribution (e.g. Goodland and Daly 1992; Weterings and Opschoor 1992). Estimates based on a scenario of halving environmental impacts while achieving equity of access to resources indicated the need for a ten-fold increase in resource productivity by the middle of the 21st century, a conclusion that led to the foundation of the International Factor 10 Club in 1993 (Factor 10 Club 1994). Others have since elaborated on this theme (e.g. Ekins and Jacobs 1994).

tion and consumption choices and the importance ascribed to them, and a restructuring of the incentives that people face when evaluating such choices. This has implications also for decisions on how technologies might be designed and how they might be used.

Shifting toward sustainability thus depends on achieving a set of interrelated changes to the economic structure, profiles of production and consumption, technologies, institutions and organisational arrangements. Eco-restructuring outcomes—evidenced, for example, in changing demands for labour and capital, changing geographies of production, consumption and environmental stress, and shifting distributions of economic and political power—will be similarly pervasive and wide-ranging (Simonis 1994). Eco-restructuring processes and outcomes can be conceived in systems terms as representing a paradigmatic shift from a pre-existing developmental model and trajectory in which low importance has been ascribed to environmental capital and benefits to ones in which the preservation of these becomes a fundamental design criterion for economic, social and environmental development. The shift can be envisaged as equivalent to another industrial or information revolution in which the long-standing *leitmotif* of past innovation processes—of seeking to increase returns to labour—is overridden by that of seeking returns to natural resources, and where the economy is reoriented from a preoccupation with maximising the values of the flows of goods produced to maintaining the values of all capital stocks (including produced goods and infrastructures) and the value of the useful services emanating from these (Weaver 1995).

Against this backdrop, what is the role of technology in eco-restructuring? Although only one element of this set of interdependent elements, it plays a pivotal role. This is because technology is the critical facilitating element, the enabling element that will make eco-restructuring technically feasible and politically viable. Unless technologies are available that, in principle, enable needs to be met without exceeding critical eco-capacities, societies, and more especially political representatives of societies, are unlikely to take measures to penalise use of unsustainable technologies. It is both sensible and politically necessary to ensure that replacement technologies are available, or could be made so in relatively short order, before established technologies and the benefits they bring are moved beyond the reach of their accustomed beneficiaries. These benefits and beneficiaries are likely to extend beyond consumption and consumers to include also the interests of investors, employees and those whose livelihood or influence is vested in established technologies. Candidate new technologies are also needed to help move the debate about the potential costs, benefits and distributional consequences of restructuring and the acceptability of these from the realm of abstract discussion and into the realm of more concrete and tangible analysis.[5]

5   Appropriate technology is a precondition for sustainable prosperity, but it is not all that is needed. Cultural and structural changes must be implemented as well as technolog-

However, this presents a 'Catch 22' to those concerned to shift development trajectories toward sustainability. On the one hand, unless incentives and framework conditions are in place that would make sustainable technologies viable—including the social, cultural and economic conditions that are needed to make a technology acceptable and profitable as well as technically feasible—there is no obvious business imperative for developing such technologies. Since the market will not reward, and may even penalise, sustainable technology innovation, sustainable technologies will not be developed. On the other hand, not having sustainable technologies 'on the shelf' is a barrier to the general restructuring of incentives,[6] which is needed if research, development and innovation are to be focused on the societal goal of sustainability. Except, perhaps, in response to an emergency, a general restructuring of incentives will be politically viable only if this has wide support in society. Before incentives can be restructured, it is necessary to be sure that an evolutionary transition to sustainability is possible that is perceived capable of delivering net benefits and is relatively painless. For this, society needs assurance that science and technology are capable of providing sustainable technologies that will meet needs in acceptable ways. To provide this assurance as well as to provide the basis for transition, some sustainable technologies need to be available 'on the shelf' ready for evaluation and implementation.

A Catch-22 phenomenon in respect to technology development is not new. To some extent, the phenomenon of new technology being locked out of the marketplace by old technology and old technology being locked in is well understood (Arthur 1989) and can be dealt with within the framework of existing institutional arrangements and without recourse to general reform and restructuring of incentives. Neither is it altogether unusual to see technology development as inseparable from the cultural and structural setting in which the technology will be used and with which it must be compatible if it is to be accepted and successful.

ical changes. Until now this 'trinity' has not been sufficiently emphasised in the whole debate surrounding sustainable prosperity. It is usual to talk only about technology or behaviour or the market and its legal framework. This gives a distorted picture. If our goal is sustainable prosperity, there needs to be a co-evolutionary development of all three: culture, structure and technology. Any debate that treats only one of these in isolation and stresses that change in any one offers a panacea is misguided. This type of debate, which is common, is by definition abstract in the extreme. Matters such as which absolutely basic needs have to be met and what risk society is prepared to take have always been debated. But it is difficult to achieve consensus on such matters if they are debated in the abstract. The danger is that abstract discussion goes round in circles. In order to move the debate on it must be made more concrete. If we can come up with a technological design or a prototype, the debate can become more meaningful. We can test the design and clarify the costs and benefits and risks. We can ask what societal and economic conditions must be met before the technology can be implemented. A technological design of this sort should not be produced by R&D scientists working alone in a laboratory; rather, it should be the result of a continuous consultation with society.

6   Such as would be implied by a reform of taxes, subsidies and government programmes and a shift toward full-cost pricing.

This basic philosophy is embedded within techniques for technology development such as Constructive Technology Assessment (CTA) that already form part of technology management practice in some situations. However, the conventional reason for lock-in and lock-out has to do mostly with the different positions of old and new technologies on their respective learning and cost-reduction curves, the cost advantage to established technologies represented by the Salter cycle,[7] the advantage to old technologies of sunk capital investment in infrastructures that is already written off and barriers to market entry raised against new technologies by organised interests vested in the old. All of these relate to the relative strength of old and new technologies in the everyday struggles that occur between them within the framework of prevailing incentives.

In respect to sustainable technologies, there is a different—more correctly, an additional—phenomenon. This is the likelihood that sustainable technologies might never be cost-competitive under prevailing incentives in distorted markets and that, to become economically viable, they ultimately depend on fundamental cultural, structural and economic reform. All markets are socially constructed and markets are subject to potential reconstruction by societies and their representatives to achieve societal objectives. Nonetheless, this dependence of sustainable technology on market reconstruction makes this a very special and intractable Catch 22. It urges new approaches for dealing with the dilemma that it represents. It also places special emphasis on finding new co-evolutionary ways of integrating dynamic and futuristic cultural and structural considerations into the dynamics of technology development. The likely views and behaviours of future generations—future citizens unable to represent themselves—must somehow be represented, considered and integrated within technology designs.

Two other aspects make this case particularly special. The first is the long lead-times involved in developing sustainable technologies that depend on such systemic or 'paradigmatic' change. Past commercial applications of fundamentally new technologies—such as television, lasers and photo-optic cables—show that the typical innovation process takes 30–50 years. Even though there is evidence that this time-span is contracting, it is unrealistic to expect it to be much less for sustainable technologies, which, as just noted, depend on paradigmatic change. Moreover, the time taken before commercialisation is sufficiently developed for a sustainable technology to make a difference to environmental stress could be even longer, since this depends on the rate of diffusion of the technology and achieving a penetration into markets sufficient to displace eco-inefficient technologies. Thus, there is no time to lose if sustainable technologies are to be ready in time to hold

---

7   The Salter cycle is a 'virtuous cycle' of self-sustaining and self-reinforcing cost reduction. As experience with a product increases, production costs reduce, which allows the price of the product to fall. This increases the demand for the product allowing economies of scale to be reaped in production and further learning, etc.

back the surge in environmental stress that population and economic growth will otherwise bring by the mid-21st century. Another implication of long lead-times is that failure to take advanced action to develop sustainable technologies could leave businesses and nations ill-placed to respond to future price changes and shortages of eco-capacity, which will not automatically be signalled by price trends in advance.

Market distortions—in the form of externalised costs, taxation structures, patterns of subsidy, exaggerated discount rates and perverse accounting procedures—already prevent price signals from reacting smoothly to developing factor shortages and surpluses. These distortions particularly affect factor markets and environmental resources. The important aspect is that there are potentially significant advantages for both nations and businesses in pre-empting resource shortages and price changes that are predictable even if not yet signalled in price trends. In the event of an increasing scarcity of eco-capacity, which world population and economic growth seem to make inevitable, costs for nations and businesses highly dependent on environmental resource use will soar. By incorporating resource scarcity into new technology designs from scratch, eco-efficiency can be achieved more cheaply and effectively than by improving existing technology incrementally. A longer lead-time should lower the cost to business of preparing responses and lead to more robust, better-designed and better-tested technologies, more attractive products, and better-prepared strategies for meeting consumer needs. With increasing scarcity of eco-capacity, eco-efficiency will become a critical element in competitiveness. The latent demand for eco-efficient goods and services will create new business opportunities and open new markets. Nations and companies best prepared will be leaders in these new markets.[8]

The second aspect lies in the contradiction that technology, while a pivotal enabling factor in sustainability, is widely mistrusted as a means of solving complex societal problems. Reaction by society against technological 'fixes' is well known, but has recently been hardened by the clear role of technology in contributing to environmental and human health problems, including those related to (un)sustainability. People are cautious about using technology to solve problems perceived as outcomes from earlier technological innovation. Past technology and past innovation processes have proved fallible. Many negative outcomes of earlier innovations were unforeseen. This is widely construed as ominous. Concern is further sensitised by the intrinsic risks and uncertainties that surround new

8   Other advantages include: for companies, the opportunity to improve standing and image among customers, employees, business associates, shareholders and local communities; for nations and companies, the possibility of reducing exposure to resource price and supply insecurities (especially those prone to forces outside their control); and, for nations, the chance to provide higher levels of environmental and risk protection for citizens and to increase the real productivity of national economies by improvements in efficiency and self-sufficiency (Weaver 1997).

technologies[9] and by people's fear of having technologies imposed upon them and becoming unknowing parties to technological 'experiments'.[10] Rapid advances in basic scientific knowledge—in energy and materials sciences, information technology, computing, flexible engineering, robotics, biotechnology, genetics, medical sciences and nanotechnology—raise further sensitivities because of the difficult moral and ethical issues to which they give rise.

THE SPECIAL SITUATION OF THE NETHERLANDS

Solving these dilemmas is important for all societies, but it is especially important for those whose production and consumption patterns impose higher-than-average demands on eco-capacity, both local and 'sequestered' from abroad. It is especially important also for societies vulnerable to the environmental and economic consequences of either a reduction in eco-capacity or reduced access to environmental resources. The Netherlands falls into both of these categories. The Netherlands is a highly developed, highly industrialised society. In relation to its small land area (3.4 million hectares), it has both a large economy (a GNP of around US$350 billion) and a large population (15 million). This gives the Netherlands one of the highest population densities in the world (around 4.5 persons per hectare) and a high income per capita (around US$23,000), providing a high material welfare. The economic status of the Netherlands is currently made possible by an intensive economy backed by high volumes of import, export, transport and value-adding industry. Much of this industry is based on adding value to imported materials and feedstocks before these are re-exported. This is the basis, for example of the petrochemicals, metallurgical and food-processing industries. Thus, for example, the Netherlands, with one of the smallest land areas of all countries, emerges as one of the most important actors in world food trade, importing low-value feeds, mostly from the Far East, and exporting high-value meat proteins, mostly to European neighbours.

The Netherlands' main port at Rotterdam is Europe's largest and busiest. The Netherlands also has one of the highest densities of motorway network in the world (5.5 km/100 km^2). Levels of both private and commercial vehicle ownership and use are high and still growing. The Netherlands has one of the world's highest per capita energy productions and consumptions (200 GJ/capita) and CO_2 emissions

9   Concern is particularly acute in respect to the human food chain and genetically engineered foods.
10   For example, if decisions are taken 'over the heads' of people or if market choices and information on these is insufficient.

per capita (9.16 tonnes/capita). Its level of pork production is among the world's highest. With 14 million pigs, the Netherlands has the world's highest density of pigs (4.1 pigs per hectare) and, also, of pig waste. These activities impose high demands on eco-capacity, both at home and sequestered from abroad. A recent study (Adriaanse *et al.* 1997) estimates the total materials requirement of the Dutch economy to be 1.275 billion tonnes per annum. This equates to 378 tonnes per hectare per year or 84 tonnes per person per year or 3.2 tonnes for each US$1,000 of GNP.

Although currently energy self-sufficient owing to natural gas production in the North Sea and the export of gas sufficient to offset its own use of imported oil, the North Sea gas fields are declining and are expected to be exhausted within the next decades. The Netherlands also faces problems with waste generation and pollution. The pig waste problem typifies a general problem: wastes and pollutants from all the value-adding activities based on imported feedstocks are left behind when the final products are exported. The small land area of the Netherlands, the flat terrain, the high water table and the density of the drainage network compound this problem. Water-borne wastes are rapidly diffused but are only slowly transported away. Furthermore, a key finding from the study of material flows is that more than 70% of the total material requirement that supports the Dutch economy arises outside the nation's borders. This implies that more than US$200 billion of presently generated Dutch GDP depends directly on using eco-capacity that may not, in the future, be available to the Dutch economy.

These statistics demonstrate that a future shortage of eco-capacity has serious implications for the Dutch economy as well as for the environment. The Netherlands already has an established role and reputation in respect to major industries and activities such as food production, chemical production, transportation and port activity. These activities are staples of the Dutch economy. Shortages of eco-capacity threaten to limit not only the output of useful goods and services from these sectors, but also the future stream of economic and social benefits that they contribute today to GNP, tax revenues, employment, export earnings and welfare. If these sectors are to continue to make comparable future contributions to the wellbeing of the Dutch economy and society, they must be restructured so that they make much more efficient use of available eco-capacity. This depends on innovation and on R&D to provide the basis for the necessary changes in technology, structure and culture. Research priorities for sustainable technology have to be determined from a broad-based definition of 'needs' that embraces a full appreciation of what is at stake in the event of future shortage of eco-capacity, including the threat that it poses to established benefits deriving from value-adding activities and sectors that are unsustainable at present. The need is not only for technological substitutes that will meet ecological criteria and maintain future flows of goods and services, but that will also maintain income, jobs, tax revenues and other indirect benefits that flow today from unsustainable technologies.

INNOVATION FOR
TECHNOLOGY RENEWAL

In summary, although not sufficient, the role of technology is pivotal for providing assurance that an acceptable transition to sustainability is feasible. But the interdependence between technological, cultural and structural change effectively blocks progress on all fronts. Strategic, systems-oriented, sustainable technologies have long lead-times and innovation horizons that are uncertain and conditional. Though needed, such technologies will not be developed automatically. Yet failure to develop sustainable technologies reduces the possibility of ever achieving transition to sustainability and of minimising transition costs in the eco-restructuring process. It also threatens the loss of opportunities that are contingent on eco-efficiency, such as to win market share and political say in the new world context that a shortage of eco-capacity and its interaction with other ongoing changes will bring. These other dynamics include globalisation, integration, liberalisation and shifts in the global distribution of economic and political power.

Sustainable technologies will need to be developed in a co-evolutionary way that integrates societal concerns into technology designs as well as building constituencies in their support and in support of the necessary restructuring of incentives and behaviours. This development must be sensitive to the mistrust of society for technological fixes, to the acute unsustainability of present arrangements, and to the need to develop technological replacements capable of fulfilling multiple roles. Beyond providing services to consumers, these roles include the positive contributions made by the technologies and activities to domestic product, the trade balance, tax revenues, employment and the transfer of skills and information across generations. Replacements for technologies that are to be phased out or scaled down on grounds of unsustainability must be capable of addressing multiple needs by fulfilling multiple functions. Such technologies will not come about through incremental improvement of existing technologies, nor will they come about without a conscious, concerted and focused effort on the part of government, business and societal groups to tackle the issue strategically and systematically. Above all, the effort depends on innovation in regard to the innovation process itself. Effectively, this forms the case for and background to the Dutch STD programme: a research and support activity aimed at strengthening innovation capacities for the development and diffusion of long-term sustainable technologies.

The remainder of this book constitutes a review and evaluation of the programme. In Chapter 2, we look in more detail at the scale of the challenge that sustainability poses for technology development and at how this challenge was translated into quantitative, operational targets appropriate for the Dutch context by Weterings and Opschoor for the STD programme. Chapter 3 looks at the past record of technology development, innovation and diffusion and draws lessons for the design of a programme aimed at strengthening innovation capacities and improving the prospects of a co-evolutionary restructuring of development pathways toward more sustainable production–consumption systems, involving sustainable technologies and sustainable use of technologies. Finally, in this opening section of the book, Chapter 4 moves from a discussion of the underpinning theory and philosophical approach of the STD programme to a description of its working method, the tools it used and its case-study illustration projects. The remainder of the book provides detailed descriptions of the case studies (Part 2) and an analysis of the programme experiences and achievements which brings out transferable lessons and recommendations for future work (Part 3).

Chapter 2
The scale of the
sustainability challenge

How big is the challenge posed by sustainable development? And how big is the challenge that it poses for technology? Answering these questions depends on invoking some operational definition for sustainability—an intellectual challenge in its own right. Precise answers to these questions are impossible to determine since answers are conditional on so many contestable assumptions. However, the absence of operational definitions is a bigger obstacle to progress toward sustainable development than is lack of precision in defining targets. Equally, procrastination that comes from contesting operational definitions and targets has been (and remains) a critical impediment to progress toward sustainability. Definitions and targets are needed now so that we can compare these against business-as-usual developments to know the broad scale of the challenge ahead. This is important, especially, when specifying the technological challenge because sustainable technology development is potentially one of the most important strategies for achieving sustainability. This is a case where it is better to be roughly right today and to take measures to steer an alternative development path than to be absolutely right tomorrow when corrective action would be too late to be effective. It is also a case where it would be better to err, if at all, on the side of over-specification.[1]

Accordingly, the STD programme commissioned a study from the Advisory Council for Research on Nature and the Environment (RMNO) to operationalise the concept of sustainable development specifically in relation to the development of technology. The report of the study (Weterings and Opschoor 1992, 1994) made a seminal contribution to the field and is now widely cited in almost all studies on

---

1 It is better to set the targets higher than is strictly necessary rather than lower, especially if we fail to meet them. The precautionary principle suggests a need to invoke operational definitions and targets that, if met, would be consistent with sustainability.

sustainability. The approach takes the concept of eco-capacity as a starting point and proposes a taxonomy of indicators based on three dimensions of unsustainability: pollution, depletion and encroachment. Derived from the concept of the environmental carrying capacity, eco-capacity refers to the bounded potential of the biosphere to supply materials and services, to absorb pollution and waste and to withstand encroachment while maintaining the capacity for processes operating within ecosystems to repair damage naturally. Taking examples of key pollutants, resources and ecosystem characteristics as illustrative cases, Weterings and Opschoor estimate the boundary conditions along each dimension, based on limiting the rate of environmental change to manageable levels and preserving stocks of key resources for future use. The result is an eco-capacity supply profile that can be compared against current and projected future eco-capacity demands. The profiles show that, even in the current situation, more eco-capacity is used than is consistent with sustainability.

Having defined and quantified the eco-capacity supply profile, albeit loosely, Weterings and Opschoor use the Ehrlich Identity to explore the challenge that this represents for sustainable development. Factoring population and income growth into the analysis, the demand for eco-capacity in 50 years' time is shown to outreach supply by a factor of 2–20 unless the path of global development is changed. Factoring greater global equity into the analysis, the estimated future mismatch in the demand and supply of eco-capacity for the richer countries is of the order of a factor of 10–50. This mismatch represents the scale of the challenge inherent in sustainable development. The study sets out three possible ways of meeting this challenge. One lies in influencing the scale and profile of the demand for final goods and services. A second lies in increasing the available eco-capacity. A third lies in radically increasing the efficiency with which eco-capacity is used. These strategies are interdependent and mutually supporting. Nonetheless, to explore the technological potential and to make best use of it, it is best to focus on the last two strategies and to take the scale of the sustainability challenge as a whole as a working hypothesis for the scale of the challenge to sustainable technology development. In this respect, the RMNO study also gave direction to the STD programme by surveying the fields where the greatest technological challenges lie: in the energy sector, in the use of non-renewable materials and in food production.

CONCEPTS OF SUSTAINABILITY

There is no uncontested operational definition of sustainability, although there is widespread agreement on the general, non-operational definition adopted by the Brundtland Commission: sustainable development is development 'that meets the needs of the present without compromising the ability of future generations to meet their own needs' (WCED 1987). The intellectual origins of this non-operational definition can be seen in proposals by Tietenberg (1984), Repetto *et al.* (1989), Solow (1986) and Pezzey (1989), which refer to the maintenance of living standards, the stream of utility or the discounted present value of utility. Dissent enters when discussion turns to the role of the environment both as a factor of production in economic processes and as a direct provider of services that contribute to life quality. This moves the discussion into difficult terrain, since it turns attention to the state of the environmental stocks and to the integrity of the functioning ecosystems that provide productive potential. Substitutability—of different forms of capital and of economically provided services for those now provided directly by nature—is the central issue of dispute.

This is a difficult issue and the debate can easily become mired in semantics and the tension of different disciplinary assumptions. Especially, a division has arisen between economists and ecologists, with the latter arguing that there are much more severe limits to substitutability than economists choose to recognise or admit (e.g. Boulding 1966; Daly 1990; Ayres 1998a). Ayres, especially, has argued that some classes of natural resources and some environmental services are essential, non-substitutable and non-replicable[2] and, also, that some are threatened[3] by today's

2   In terms of resources and services that are essential, non-substitutable and non-replicable, Ayres gives, as examples, the need for arable land, water and a benign climate in agriculture, for the action of reducing bacteria in nutrient recycling and for stratospheric ozone in protecting life forms from harmful ultraviolet radiation. More generally, his definition of sustainability precludes, among others, major climatic change, species loss, deforestation, desertification and accumulation of toxic heavy metals and non-biodegradable halogenated organics in soils and sediments (Ayres 1998a: 8). All of these are potentially threatened by four trends: continuing population growth, the raw material demands of industrial activity, the emission of wastes and pollution from economic activities and anthropogenic disturbance of balanced environmental systems. Ayres divides the discussion of these between non-controversial issues, in which he includes population, resources and technology, and controversial issues, in which he includes pollution, productivity and biospheric stability. Ayres's major concern is in respect to the stability of the biosphere and threats to this.

3   Ayres specifically cites the dependence of life on the Earth's nutrient recycling system, which secures a supply of nutrients in forms and amounts greater than would be found in the Earth's crust or the prebiotic atmosphere or hydrosphere (Ayres 1998a: 14). His argument develops from the fact that the Earth is a complex, interactive, self-organising system, which is held in a stable state far from its thermodynamic equilibrium condition. The system maintains an orderly state by capturing and using radiant solar energy. The energy is captured by living organisms and passed along the food chain. As well as fixing and moving energy, this secures the transformation and movement of materials and the performance of essential functions, such as nutrient cycling. The biosphere is therefore an active element in the Earth system. Because the system is complex and non-linear, its dynamic behaviour is potentially chaotic. The stability of the system, which depends on complex feedback loops, is only assured when the system-state remains within a certain range. The resilience of the system—its tendency to remain within its original domain—is indeterminable. Ayres's main thesis is that the grand nutrient cycles are part of the Earth's regulatory metabolism. Moreover, pollution of natural systems with xenobiotic

economic activities and development pathways (e.g. Ayres and Kneese 1971; Ayres 1978, 1998a). As several others before him, Ayres argues that there must be ecological limits to economic activities and he moves the discussion along by trying to specify which important environmental qualities should be preserved and why. For example, he specifies that there must be limits to the permissible increase in concentrations of greenhouse gases in the atmosphere, hydrogen ions (acids) in surface waters and soils, nitrates and nitrites in groundwater and toxic heavy metals in soils and sediments as well as to the permissible losses of topsoil, groundwater, tropical forests, important ecological habitats and biodiversity.[4] Nonetheless, aware of the dangers involved, he stops short of setting out quantitative limits.

In spite of the dangers, it is important to quantify these limits, however imperfectly, and for this it is necessary to adopt some operating principles and assumptions. The RMNO study is based on the assumption that the biosphere offers global society a finite means of support in the form of stocks of natural resources, supplies of essential services and resilience to depletion, pollution and encroachment. It seeks to anticipate the scale of impact on the environment if present demographic, economic, social and technological trends worldwide continue unchecked and policies remain as now. The time-horizon is to the year 2040. Since much depends on the resilience of nature and on the ingenuity of human society to develop and implement new solutions, the exercise accepts that a number of different positions could be assumed in attempts to demarcate eco-capacity, ranging from *laissez-faire* to cautious (Opschoor and van der Ploeg 1991). The study could have assumed that nature is resilient and/or that knowledge will develop fast enough to yield new solutions to any resource shortages or environmental changes. However, from what is known today about environmental, economic and technological processes, a cautious, risk-averse position is more prudent. It is safer to assume that all environmental resources and services that are needed today will be needed in the future and at levels implied by current trends in use, rather than to assume that some may not be needed. Adopting a prudent stance means that targets will be set that challenge society to develop the necessary knowledge and expertise to avoid resource shortages and that quantify the scale of the task that this implies.

---

substances or natural substances in unnatural concentrations creates imbalance analogous to the effect that toxic substances would have on the metabolic processes of living organisms. The disturbance has an indeterminable potential to destabilise the Earth system and to shift it to a new thermodynamic condition.

4   Ayres translates these conditions into three operational rules and targets for development into the mid-21st century. Use of fossil fuels must decline to very low levels. Agricultural, forestry and fishery practices must be radically overhauled, with less dependence on chemicals and mechanisation. Net emissions to the environment of long-lived toxic chemical compounds, such as heavy metals and halogenated organic compounds, must drop to near zero levels (Ayres 1998a: 32).

DEFINING A SUSTAINABLE
LEVEL OF USE

For the three dimensions of environmental impact—depletion, pollution and encroachment—the study invoked a hierarchy of sustainability criteria, ranging from the general to the specific. For depletion, the general sustainability criterion is that there should be no absolute exhaustion. For pollution, it is that there should be no accumulation of polluting substances or any lasting effects for coming generations. For encroachment, the general condition is that rates of loss must not exceed rates of restoration or replenishment by natural or artificial means. In each case, more specific criteria can be worked out only for illustrative resources, pollutants or damages and in respect to the relevant scale of the impacts—global, continental or national. In the case of depletion of mineral resources, the basic reference is the proven or economically recoverable global stock of specific minerals. For renewable resources, the relevant reference is the rate of natural regeneration. For pollution and encroachment, reference evidence comes from studies of natural processes and norms. Evidence on natural tolerances to pollution by specific substances comes from studies of natural loadings, buffering capacities, substance concentrations and variability in these. For encroachment, appropriate references derive from studies of natural rates of formation and loss of critical ecosystem components, such as soil and groundwater, and from the biodiversity found in undisturbed natural ecosystems.

For depletion, the RMNO study collated evidence on the stock of non-renewable resources for a set of indicative fossil fuels (coal, oil and gas) and metals (aluminium, copper and uranium). It translated these into eco-capacity limits using the criteria that stocks should remain at, or be brought up to, a level sufficient to allow a further 50 years' use at prevailing use rates. In aggregate, the total global stock of fossil fuels is enormous, somewhere between 230,000 EJ and 260,000 EJ. This is sufficient for several centuries of use at current usage levels. However, the economically recoverable stock is much smaller (less than 36,000 EJ or sufficient for 100 years at current rates of use) and coal, the most polluting fossil fuel, makes up three-quarters of this. Both the total reserves of oil and gas (9,000 EJ and 9,900 EJ respectively) and the economically recoverable reserves (6,000 EJ and 4,500 EJ) are small in absolute terms and in relation to current levels of use (125 EJ and 60 EJ). At current use rates, the economically recoverable reserve of oil is within 50 years of exhaustion; the reserve of natural gas will last only 75 years. Rates of use are, however, growing rapidly, so the stocks may not last even this long. Moreover, whereas the quantity of energy is preserved in all transformations, its quality (exergy) is not preserved. The stock of fossil energy is therefore limited in a way that other mineral resources, which in principle can be recovered and recycled, are not. Another important aspect is that the number of primary energy sources is highly constrained, which limits the substitution possibilities among the different mineral energies.

For metals, the situation varies widely depending on the specific material, its uses and substitution possibilities. Data was collected on the economically recoverable stocks of three indicative metals. Global reserves of aluminium (5,750 million tonnes) are capable of supporting over a century of continuing use at current usage rates (35 million tonnes per annum), without recovery or recycling. In comparison, the global copper reserve (500 million tonnes) is small compared to the annual use rate (15 million tonnes) and already below the 50-year criterion. The total uranium reserve (8–30 million tonnes) is sufficient to provide for 130 years' continuing use at the present usage rate. Again, rates of use of these metals are rising. In the case of uranium, a main application is for energy generation. Future extraction rates will depend on the energy demand, which is growing, the availability of substitute primary energy sources of whatever origin and the rate of uranium recovery and recycling. In the case of all metals and inorganic materials, recovery and recycling offer an alternative to drawing on primary reserves, but they require energy and consequently also give rise to environmental impacts.

In the case of non-renewable resources, the study assumed that human use should be small in relation to the regeneration rates in undisturbed natural systems. This criterion was quantified for global terrestrial primary productivity, global terrestrial animal biomass and the natural rates of species loss. Global terrestrial primary production is estimated to be 2,400 EJ per annum. Global terrestrial animal biomass (dry weight) is estimated to be 1,300 million tonnes. The RMNO study used work tracing the growth in the human claim on both terrestrial primary productivity (Wright 1990) and animal biomass (Westing 1981) to identify periods when human usage appeared not to give rise to disruptive environmental impacts. This level of use—in each case equivalent to 20% of the regeneration rate—was taken as indicative of the maximum sustainable level of claim. This is 480 EJ in the case of terrestrial primary productivity and 260 million tonnes for animal biomass. For biodiversity, it is more difficult to estimate with any accuracy either the number of species extant or the natural rate of species introduction and extinction. Estimates of the number of species on Earth range from 4 million to 30 million. The average natural extinction rate is perhaps one to five species per decade (IUCN 1983; Wilson 1989), although it is clear that there have been five episodes of substantial species extinction in the geological past. The RMNO study assumed that a sustainable extinction rate should not exceed the average natural rate by very much. It took a figure of five species per annum as the maximum sustainable level.

The general criterion for pollution—that there should be no accumulation of polluting substances or any lasting effects for following generations—was quantified for global emissions of the greenhouse gas carbon dioxide, acid deposition on the continental scale, the deposition of nutrients and the loading of Dutch surface waters with cadmium, copper, lead and zinc. The notion of sustainable level in regard to pollution is predicated on the idea that natural and artificial processes can compensate for some of the damage and that some level of loading is normal in natural systems. Pollution therefore concerns only xenobiotic substances and natural substances in unnatural concentrations.

▷ For greenhouse gas accumulation, the RMNO study took the IPCC guidelines on the maximum permissible rate of temperature increase (0.1°C per decade) and ultimate temperature ceiling (a total increase of 2.0°C) and adopted the sustainability criterion developed from these by Krause for annual permissible levels of carbon release over the next century. This is 2.6 gigatonnes of carbon per annum (Krause *et al.* 1990).

▷ The criterion for acidifying substances, including sulphur dioxide (SO_2), oxides of nitrogen (NO_x) and ammonia (NH_3), was taken from work on critical loads, which translates these different sources into units of acid-equivalence and compares anthropogenic deposition with natural loads and buffering capacities (Nilsson and Grennfeldt 1988; Heij and Schneider 1990; Hetteling *et al.* 1991). Since critical loads and damage potential depend on site-specific factors, it is possible to develop only a generalised sustainable level based on average conditions. The average annual deposition from natural sources is 200 acid-equivalent units per hectare. On the basis that the sustainable level should be not much greater than this, the RMNO study took 400 acid-equivalent units per hectare per year as the sustainable level.

▷ Similarly, the sustainable level for eutrophying substances (nitrogen and phosphorus compounds) that can be added to soil and water were based on the quantities of nutrients that are annually removed in crops or are released to the atmosphere by crops during growing processes and to the maximum permissible concentrations of residues in run-off and drainage water. For phosphorus compounds, the RMNO study adopted 30 kilograms per hectare per year as the sustainable level of deposition. For nitrogen, the relevant figure is 267 kilograms per hectare per year.

▷ In regard to toxification by heavy metals, where concern is for substances that occur naturally only in very small quantities, the RMNO study concludes that the sustainable level should be based on ensuring that the impact on the Earth from anthropogenic deposition should be small in relation to natural flows. Because heavy metals attach themselves to sediments, natural concentrations in surface waters can be estimated by examining undisturbed, pre-industrial sludge deposits. The RMNO study took guidance from studies of four metals, adopting an annual release to surface water of 2 tonnes for cadmium, 58 tonnes for lead, 70 tonnes for copper and 215 tonnes for zinc.

For encroachment or disturbance of natural systems, the RMNO study looked both at soil erosion and groundwater loss. These might also be considered in the depletion category. For this reason, in a later version of their report, Weterings and Opschoor interpreted encroachment as disturbance of natural system integrity and function (Weterings and Opschoor 1994). Even in the 1992 report, however, the best indicator found to describe the effect of groundwater loss on wetland ecosystems was reduction in species diversity in specific ecotope groups arising from dehydration effects on the incidence and completeness of the natural biodiversity. The sustainable level was therefore defined in terms of a maximum permissible percentage reduction from the reference biodiversity for each of several different natural wetland ecotope groups. These levels, ranging from 75% to 95%, are based on work by Claesson, who suggests that, if achieved, these targets would guarantee a sustainable development of wet and damp terrestrial nature in the Netherlands (Claesson 1991). For soil erosion, salinification and impoverishment, the reference was taken from work on the natural rate of soil erosion and replenishment (Judson 1986), which estimates this to be 9.3 billion tonnes per annum.

PROJECTING FUTURE LEVELS OF IMPACT

On each eco-capacity indicator, the sustainable level of use should be compared with the projected level of demand in 2040 on the assumption that present social, economic and technological trends will continue and that policies will remain unchanged. For this, the authors of the RMNO study prepared scenarios about social developments based on extrapolating existing trends in population and economic growth to provide an overview of the general scale of the challenge that limited eco-capacity poses for sustainable development. Evidence was also drawn from detailed studies of anticipated future trends in resource use, pollution and encroachment in order to paint a more detailed picture showing which environmental resources and services are most at risk and which economic activities give rise to the greatest and least sustainable eco-capacity claims.

Two different population forecasts were developed (high and low) based on different scenarios about population growth rates over the forecasting period. The low forecast assumes that the present population growth rate of the Northern countries (0.6% per annum) will reduce to 0.5% over the period 2000–2025 and that, thereafter, the population will remain at a stable level (zero growth rate). The low forecast for the Southern countries assumes that the present growth rate (2.0%

per annum) will drop first to 1.5% over the period 2000–2025 and thereafter to 1.0% per annum. This low forecast gives a world population of 9.4 billion in 2040, with a population of 1.3 billion in the North and 8.1 billion in the South. The high forecast assumes that growth rates in the two regions remain at their present levels throughout the entire forecasting period. This gives a world population of 12.7 billion in 2040, with a population of 1.4 billion in the North and 11.3 billion in the South. Both forecasts therefore reveal a sharp increase in the population of the South and a virtually constant population in the North. Equally, both reveal a very large absolute increase in the world population. Even the low scenario estimates a doubling by 2040 relative to the present world population and the high scenario anticipates growth to two-and-a-half times the present population size.

Similarly, the study made a range of estimates of the future average level of per capita income based on assumptions about rates of economic growth in the North and South. The future level of wealth was estimated by assuming various annual economic growth percentages and dividing the total GNP across the expected population size. In the 1980s the North experienced an annual economic growth rate of 3%. Continued until 2040, such a growth rate would lead to a quadrupling of per capita income. Even a more modest growth rate of 2.5% per annum would lead, by 2040, to an average per capita income in the North equal to one-and-a-half times that of the USA in the late 1980s. In contrast, because of the expected population growth in the South, substantial rates of economic growth would be needed even to achieve modest increases in per capita income. An annual growth rate of 3.5% over the period to 2040 would leave the South with an average per capita income only one-third of the level enjoyed in the North in the mid-1980s. An annual economic growth of 5.5% (low population forecast) or 6.5% (high population forecast) would be needed for the South to attain a per capita income equal to that of the Northern countries in the mid-1980s. This is considered to be the minimum level necessary to eliminate the worst forms of poverty (WCED 1987).

The RMNO study also drew attention to the existing disparities in wealth between the two regions and how these might grow under different development scenarios. On average, per capita income in the North is currently a factor of 10 higher than in the South. If we assume the lower economic growth forecast for the North (2.5% growth per annum) and an economic growth rate of 3.5% per annum in the South, the present ten-fold gap in per capita income between North and South will not be reduced, let alone closed, by 2040. In the case of 5.5% annual economic growth in the South, the income gap would reduce to a factor of 3. However, a minimum annual economic growth rate of 7.5% per annum in the South is needed to bridge the income gap with the North completely by 2040. Even with conservative assumptions, the RMNO study concluded that the combined effects of population growth and growth in per capita income will lead to drastic increases in the use of eco-capacity by 2040 if wealth creation is based on current production and consumption methods.

The broad challenge that this presents for sustainable development is clear from the Ehrlich model (I = P.C.T), which illustrates the relationship between the aggregate claim on eco-capacity (I) and the aggregate scale of economic activities (P.C) as a function of average eco-efficiency (T). The model highlights that projected demographic and economic developments underlie the need for drastic parallel developments in regard to all strategies for achieving sustainable development, and, especially, eco-efficiency. The scale of the challenge to technology is a function, also, of progress in improving equity. If the ten-fold gap in per capita income between North and South remains, the environmental claim per unit of wealth will have to be reduced by at least 80% over the 1990 claim just to hold the aggregate claim on eco-capacity at the 1990 level. This equates to a required eco-efficiency improvement of at least a factor of 5. If we wish, also, to achieve an equal distribution of wealth between the North and South, the needed reduction will be in the order of 95%. The required eco-efficiency improvement is of the order of at least a factor of 20. As mentioned, the 1990 draw on eco-capacity is not necessarily sustainable and larger improvements may be needed to reduce the aggregate claim to below the 1990 level.

The RMNO study sought to quantify the scale of the challenge in respect to each of its selected eco-capacity indicators and in relation to quantitative sustainable claim levels. For each indicator, the study authors projected the development in the claim on eco-capacity to 2040 on the assumption that social, economic and technological trends will continue and policies will remain unchanged or took the projections of other, more detailed, studies in respect to specific indicators. For example, the authors drew on detailed energy consumption forecasts made by Kouffeld (1990) and by Holdren and Pachauri (1991) and on projections of the human use of terrestrial primary productivity made by Wright (1990). In respect to carbon emissions, the study authors drew on the Intergovernmental Panel on Climate Change (IPCC) 'business-as-usual' scenario, which suggests that emissions will have increased by 2040 to about 13 gigatonnes of carbon per annum. For acidification, the RIVM (Netherlands Ministry of Housing, Physical Planning and the Environment) forecast for Europe was used, which anticipates an average deposition of 2,400–3,600 acid-equivalents per hectare per annum by 2040 (RIVM 1991). Only in three cases was it found impossible to estimate the claim on eco-capacity in 2040. The missing indicators are for uranium consumption, eutrophication and dehydration. For all the other cases, it was possible to specify a quantitative indicator. The outcome from this exercise is presented in Table 2.1.

Table 2.1 compares the expected claim on eco-capacity in 2040 with the estimated sustainable usage level across the set of different indicators. For example, on the basis of population and economic growth forecasts and expectations about the future structure and level of electricity generation and farming activities in Europe, the level of acidifying emissions by 2040 is expected to be at least 2,400 acid-equivalents per hectare per year. This compares with an estimated sustain-

INDICATOR	SUSTAINABLE LEVEL	EXPECTED LEVEL (2040)	REDUCTION REQUIRED (%)	SCALE
Depletion				
Depletion of fossil fuels				
Oil	50-year criterion	Stock exhausted	85	Global
Gas	50-year criterion	Stock exhausted	70	Global
Coal	50-year criterion	Stock exhausted	20	Global
Depletion of metals				
Aluminium	50-year criterion	Stock >50-years	none	Global
Copper	50-year criterion	Stock exhausted	80	Global
Uranium	50-year criterion	Depends on use of nuclear energy	Not quantifiable	Global
Depletion of renewable resources				
Primary productivity	20% of total	50% of total	60	Global
Animal biomass	20% of total	50% of total	60	Global
Biodiversity	5 species lost per annum	365+ species per annum	99	Global
Pollution				
Emission of carbon	2.6 gigatonnes per annum	13 gigatonnes per annum	80	Global
Acid deposition	400 acid-equivalent/ha/yr	2,400–3,600 acid-equivalent/ha/yr	85	Continental
Nutrient deposition				
Phosphorus	30 kg/ha/yr	No quantitative data	Not quantifiable	National
Nitrogen	267 kg/ha/yr	No quantitative data	Not quantifiable	National
Heavy metal deposition				
Cadmium	2 tonnes per annum	50 tonnes per annum	95	National
Copper	70 tonnes per annum	830 tonnes per annum	90	National
Lead	58 tonnes per annum	700 tonnes per annum	90	National
Zinc	215 tonnes per annum	5,190 tonnes per annum	95	National
Encroachment				
Dehydration	Reference year 1950	No quantitative data	Not quantifiable	National
Soil loss	9.3 billion tonnes per annum	45–60 billion tonnes per annum	85	Global

Table 2.1 **Projected versus sustainable claims on eco-capacity**

Source: Weterings and Opschoor 1992

able level of only 400 units. Similarly, deposition of cadmium in Dutch surface water is expected to grow to 50 tonnes per annum by 2040 if present trends continue, whereas a sustainable level of deposition is only 2 tonnes per annum. Global soil erosion is expected to result in the loss of at least 45 billion tonnes of topsoil per annum by 2040, compared with the sustainable level of just over 9 billion tonnes. While the report authors say that there is no consensus among scientists on these estimates, the estimates are based on insights from the most up-to-date, authoritative work available, and thus represent a solid indication of the order of the needed reductions in environmental impact.

Clearly, Table 2.1 shows that, in the event of trends and policies remaining as now, the level of environmental impact in 2040 will exceed eco-capacity in many respects. In the case of most indicators, the level of impact is of the order of several times the sustainable level and reductions of 50%–95% are desirable relative to the expected claims, which implies that trends need to be deflected drastically. Nonetheless, even Table 2.1 does not fully describe the potential scale of the challenge that limited eco-capacity represents for the North, since it concentrates only on the aggregate situation. It does not reflect the implications that would arise if available eco-capacity had to be shared more equally among world citizens in the future. The significance of this becomes clear when we consider how unevenly the use of eco-capacity is distributed today. One-fifth of the world population lives in the North. However, the North takes up 25% of globally used biomass, 80% of globally consumed energy and 90% of globally consumed metals. It accounts for 74% of carbon emissions. For reasons of equity and greater international security, some redistribution in the use of eco-capacity will be necessary alongside the absolute reductions called for in Table 2.1. The North may have to relinquish some of its claim on the limited eco-capacity that is available to enable development in the South.

The differential population growth rates of the North and South add to the challenge. By 2040, the population of the North will constitute only one-seventh of the world population. If the present split in the use of global eco-capacity were maintained, the faster population growth in the South would mean that the gap in per capita use of eco-capacity, which is already highly inequitable, would grow between the North and South. However, if use of available eco-capacity were to be distributed evenly across world citizens at the same time as the overall claim on eco-capacity is reduced to a sustainable level, the challenge that living within eco-capacity constraints implies for the North takes on a new meaning. Taking carbon emissions as an illustration, the North accounts currently for 74% of total global fossil carbon emissions. Assuming that the North and South keep their relative shares of eco-capacity at the current level, the North would have to reduce its emissions of fossil carbon by a factor of 5 compared to its expected emissions level in 2040. However, if use of the available eco-capacity were evenly shared, the North would have to reduce its emissions by a factor of 25. The sustainability challenge is commensurately greater for the North under this equity criterion across all indicators and is often in the order of a factor of 10–50: for example, oil consumption (factor 40), copper consumption (factor 30) and acid deposition (factor 50).

TECHNOLOGICAL CHALLENGES AND OPPORTUNITIES

What does this analysis say about where the major technological challenges and solution directions lie? In their report, Weterings and Opschoor stress that, whereas their work focused on individual impact indicators, the different indicators are strongly related. The relationships, such as between depletion and pollution, are important for describing generic technological solutions. Especially, there are relationships between depletion of fossil fuels and emissions of carbon dioxide, heavy metals and acid-forming compounds. There are relationships between the depletion of stocks of heavy metals and environmental toxification. And there are relationships between the use of biomass and the loss of biodiversity, eutrophication, acidification, soil erosion and ecosystem dehydration. Three technology clusters in the areas of energy services, industrial materials supply and human nutrition account today for much of the total environmental stress. By implication, efforts should focus on finding new solutions to providing services and meeting needs in these fields.

In the field of energy technology, the main problems arise from the use of fossil fuels, which today simultaneously gives rise to both depletion and pollution impacts. This raises the question of which shifts in technological developments are needed in the energy sector to reduce energy consumption per capita and to bring about far-reaching reductions in the quantities of fossil fuel consumed and CO_2 emitted per energy service unit. Possible solution strategies lie in developing renewable energy forms, improving efficiency across source-to-service energy chains, dematerialisation, immobilisation of wastes and nature development and conservation. In the field of industrial materials, where the major problems arise with depletion of mineral stocks and pollution, the major strategies for improvement lie in finding renewable substitutes, dematerialisation, recycling, immobilisation, nature conservation and recycling. In the field of food production and human nutrition, the major problems arise from use of eco-capacity for arable farming and intensive livestock production. Ploughing and land clearance is giving rise to soil erosion. Irrigation is making inroads into groundwater stocks and causing dehydration. Intensive food production is leading to the release and accumulation of acidifying, eutrophying and toxic compounds. Solutions lie in shifting consumption patterns and lowering environmental impact for each unit of food that is produced and consumed. Nature development and conservation based on more integrated food production systems, dematerialisation of food production processes, nutrient recycling and immobilisation of residues all offer potential solution directions.

CONCLUSIONS AND RECOMMENDATIONS

The main contributions of the RMNO study were to quantify the limits to eco-capacity and to explore the scale of the challenge that living within eco-capacity limits and sharing eco-capacity imply. The clear conclusion is that eco-capacity represents a major challenge for sustainable development. Today sustainable development is mainly interpreted in terms of improving the environmental performance of our existing ways of creating wealth. But the scale of the challenge—ten-, twenty- and even fifty-fold reductions—is clear evidence that today's approaches are not enough. Sustainable development needs to be inter-preted as a strategic challenge. Solutions are needed that break existing trends in current development processes. There are at least three main strategies for this— influencing strategies, extension strategies and eco-efficiency strategies—all of which should be explored. A clear need is to influence the processes of techno-logical innovation to give a stronger focus on long-term issues. There is a need to explore the contribution that sustainable technology development could make and, equally, to ascertain what structural, social and cultural changes are required alongside trend breaks in technology development to bring about a sustainable future. A prudent working hypothesis is for the North to target ten- to fifty-fold reductions over the anticipated claims on eco-capacity by the middle of the 21st century.

Chapter 3
Technology and technological change

The aim of this chapter is to review studies of technological change for insights that they provide into how a public programme aimed at stimulating the development of sustainable technologies might be designed. In the process, we try to establish what we mean by technology and innovation. We look at patterns and processes that are revealed in the record of past technological change and at the co-evolutionary relationship between technological, economic, social and environmental change. The chapter begins by looking at individual technologies and the technology life-cycle. It then moves on to consider relationships between technologies and, especially, the emergence of clusters or bundles of related technologies. Such technology clusters are associated with major shifts in the techno-economic paradigm and in overall development trajectories. Given the difficulties of technology forecasting and the poor past record of public policy attempts to influence technological trajectories, the chapter also includes a short discussion of the role of government in this area. We conclude that there is a case for involvement and, indeed, that government involvement is absolutely necessary to try to steer technological change toward the achievement of public policy goals. But a public programme needs to be well designed if it is to have any chance of being effective. The role of government should be to try to influence technological and development trajectories indirectly, by facilitating innovation in respect to the innovation process itself, rather than by picking and backing prospective technological 'winners'. At the end of the chapter, we set out some generalisations about technological change processes and draw lessons that hold important implications for the STD programme design.

WHAT IS
TECHNOLOGY?

It is useful to begin by defining some terms and concepts that are commonly used to describe and analyse technological change. To start, we need a definition of technology itself. We also need to distinguish the different stages in the process of technology development and, especially, invention, innovation and diffusion. In addition, it is useful to define such concepts as 'technological limits', 'technological barriers', 'technological distance' and 'breakthroughs' because these are widely used in describing technological change processes. Particularly useful is the 'life-cycle' metaphor, which describes the evolution of technologies over time. There is an innovation life-cycle, which describes how the technical performance of a particular technological solution improves over time. There is also an associated diffusion life-cycle, which describes how a particular technological solution is taken up by society and how it builds its market over time. The life-cycle metaphor is used also in relation to whole clusters of technologies and to techno-economic systems that define entire phases of development.

Many different definitions of technology have been proposed. Most technology analysts define technology in relation to function. Grübler, for example, says that technology is what enables humans to extend their capabilities and to accomplish tasks that they could not perform otherwise (Grübler 1992). Ayres is more specific about the nature of these tasks: technology can be regarded as knowledge combined with appropriate means to transform materials, carriers of energy or other types of information from less desirable to more desirable forms (Ayres 1994). As such, technology is a resource that enables us to make other resources productive. Some definitions of technology focus narrowly only on the most tangible elements of technological solutions, the physical artefacts or technological 'hardware'. These include such things as process equipment, infrastructures and products. Other definitions are more broadly based and include, also, notions of technological 'software'. Software represents the knowledge, know-how, practices and organisational skills needed to develop technologies and to produce and use artefacts. Although not so tangible as technological hardware, software is reflected in organisational and institutional arrangements, which are as important as hardware in determining the capacity of societies to perform useful tasks and how societies increase these capacities by pushing out the technological boundaries imposed by the limits of the current hardware and software. They directly influence the inception, development and diffusion processes of technologies.

The historical record shows that the process of development is episodic and that output and productivity growth in different periods have been stimulated by changing structures of economic activities and technological change associated with clusters of co-related and co-evolving technologies, organisational arrangements and institutions. These clusters describe entire social, technological and economic systems of production and consumption, which are dynamic and evolving but which have coherence through mutually reinforcing feedback relationships between and among system components. In this broad sense, the whole system configuration or techno-economic paradigm that characterises a particular phase of development (Freeman and Perez 1988) can be considered to constitute 'technology' since it is made up of technological hardware and software in the guise of artefacts, know-how, organisational arrangements and institutions and provides the context in which technology is used, produced and embedded (Kline 1985). Such techno-economic systems and their related periods of development are often named after a dominant physical artefact or symbol, such as 'the age of iron and steel' or 'the steam era', but this is to downplay the inseparability of the various elements of the system and the relationships among them.

INVENTION

The process by which new technological solutions are discovered or 'invented' is not well understood. Some important discoveries have been made accidentally or incidentally, which may give the impression that this is a random process. However, the historical record shows that 'inventiveness' is unevenly distributed in time and space and across societies. There have been golden eras often associated with discoveries in basic science that paved the way for concentrated bursts of creative invention as the new knowledge is applied to long-standing problems and leads to the discovery of new ways of performing tasks and meeting needs. The discovery of electricity, for example, paved the way for a raft of new solutions to old technological challenges, such as providing illumination, communication, entertainment, power, heat and similar services. It also made entirely new services possible, such as sound recording and reproduction, and therefore stimulated new demands where none previously existed, such as for radio and television. Building on the discovery of electricity, later basic research, such as in the area of semiconductivity, paved the way for a revolution in information technology. Recent discoveries of the photoelectric properties of some materials are continuing this 'cumulative' trend of knowledge creation and invention. The same holds for recent basic discoveries in such fields as genetics and nanotechnology.

The build-up of knowledge over time can also help to make the search for technological solutions more systematic and efficient. Patterns and relationships can guide scientists and inventors to search only those domains most likely to yield solutions. The status of basic science is, therefore, an important factor in inventiveness. Although there is no fixed relationship between the two, it is clear that, without breakthroughs and discoveries in basic science, the scope for inventiveness diminishes. The investment in research, both basic and applied, the balance between the two and the allocations across different research fields, the number of recent discoveries in basic science and their significance *vis-à-vis* the technological needs and challenges of the day are, therefore, all contributory factors to inventiveness. But there are other, less tangible factors, too. The strength of the innovation culture within a society or enterprise is clearly important, as, also, is the strength of supporting institutional and organisational infrastructures. At different times, creative leadership has shifted between societies. Although there is no fixed relation between the two, economic leadership has often changed with it.

In effect, technological creativity is independent neither of effort nor context. It requires both, is influenced by both and, in turn, influences both. The seminal contributions of Veblen (1904, 1921, 1953) and Schumpeter (1911, 1934) to our understanding of the processes underlying technological creativity stress the interrelationship between the social and economic context and technological change and the importance of interactions among networks of social actors in the development and improvement of technological solutions. Contrary to what is often supposed or is tacitly assumed, new technologies are not generally the products of a moment of inspired creativity on the part of inventors, designers and engineers working in remote laboratories set apart from the disturbances of everyday life. Rather, technologies are 'socially constructed' (Smith and Marx 1994). Social and economic values, problems, opportunities, expectations and incentives influence the creative process. New solutions are the result of deliberate R&D expenditures and efforts and the product of a constructive dialogue between many different social actors, technology developers and users especially. Technology is therefore endogenous to the society and economy. Society and the economy shape it and, in turn, it is a factor in shaping social and economic conditions. The interactions are two-way and technology, therefore, is most correctly conceptualised as an integral element of a techno-economic system characterised by co-evolutionary dynamics.

Another little-understood process is technology selection. Typically, many candidate technological solutions are proposed for accomplishing the same task. Often these are designed to different specifications, sometimes reflecting different criteria or different weights attached to similar criteria, or representing different technical approaches to a similarly specified design problem. It is not always the case that the technically superior or most cost-effective solution is selected for further development and commercial introduction by its initiators or is successful in competition with others after release onto markets. Sometimes, inferior designs are locked in and superior ones are locked out for historical reasons. The 'qwerty' keyboard layout, still used today for electronic typesetting but designed originally to stop the keys of mechanical typewriters from entangling, is the oft-quoted example of 'lock-in' (Arthur 1989). Although typesetting technology is now several generations removed from mechanical keyboards, the fact that a generation of stenographers learned to touch-type using 'qwerty' keyboards was an impediment to changing the layout even after there was no technical need to retain it. Uneven market power and market distortions are also factors in technology selection.[1]

INNOVATION

Innovation refers to the process of achieving technological improvement. Broadly, this is of two types: Schumpeterian and Usherian. Schumpeterian innovation is radical and trend-breaking. It refers to the achievement of a technological break-through and its first commercial introduction as an entirely new technological solution. The breakthrough may take the form of an entirely new process, product, form of organisation or market. The new breakthrough will have a genealogy traceable to earlier discoveries. Nonetheless, the essence of a radical breakthrough is that it represents a fundamentally new approach that departs from pre-existing engineering practice and technologies and is not a continuous development of any single former approach. To illustrate this point, Schumpeter used the example of the railroad, which is not an evolutionary development from the horse-drawn carriage, the former dominant transport technology, but a new solution.[2] 'Add successively as many mailcoaches as you please; you will never get a railroad thereby' (Schumpeter 1934: 64). In a single, discrete event, the introduction funda-mentally changes the relationships between economic inputs and outputs or the nature of the constraints under which the relationship between inputs and outputs can evolve (Grübler 1998). This causes a quantitative or qualitative jump in productivity or performance. The essence of Schumpeterian innovation, then, is this productivity discontinuity, the disequilibrium it introduces into the economy and the economic restructuring it initiates (Fourastié 1949).

Usherian innovation, in contrast, is gradual and incremental. It refers to the process of optimising the technical and economic performance of an existing solution following its introduction, which is driven by competitive pressures and is facilitated by the learning process that occurs through cumulative experience in producing and marketing the technology. It is helped, especially, by market infor-mation from users that is fed back to producers and by a virtuous cycle between

---

1   The success of the PC over superior computing alternatives has been attributed to the fact that IBM, an important early promoter of PC technology, already had an established worldwide network of outlets capable of providing local support and back-up service to customers. The success of VHS-format video players over several superior designs has been attributed to video-film distributors who used this format, making it in most cases the only choice for consumers who wished to view pre-recorded films.

2   In the case of the railroad, this was based on combining two earlier developments—the static steam engine and the railtracks developed for horse-drawn railways—into a single new technological approach to transportation: the steam railway.

product improvement and cost reduction on the one hand and increases in sales on the other. Empirical studies show fairly consistent findings about the relationships involved. A study of over 100 technologies revealed cost reductions of 10%–25% with each doubling of cumulative output (Argote and Epple 1990).

Innovation tends to be a cyclical process and the two forms of innovation—Schumpeterian and Usherian—are complementary. Phases of incremental improvement are punctuated with short periods of radical innovation. A radical breakthrough technology is often a response to stagnation in the incremental improvement process of an earlier solution as its performance limits are approached, to problems associated with the further use of an older solution and to creativity following new discoveries in basic science. After introduction, the new solution is then improved over time. The time-path of improvement tends to follow an S-shaped curve, with a period of gradually accelerating improvement on the original design, a phase of very rapid improvement as learning and experience with the solution accumulate and, then, a slowing down of improvement as the limits of the new technology are gradually reached.

It is worth spending a little time discussing the nature of these 'limits', since they are characteristic of all technological solutions. Analysts tend to distinguish two types of limit. One is set by the existence of physical and chemical laws that specify absolute maximum potentials for qualities such as the conversion efficiencies of processes. For example, an absolute minimum quantity of carbon is needed to reduce iron ore to pig iron, giving an absolute benchmark for measuring the efficiency of blast furnace smelting processes. No technology can improve on the theoretical maximum efficiency. However, there are also configuration-dependent and context-dependent technological limits, which in practice are much more pervasive and constraining than any absolute limits. For example, the applications context for a solution usually specifies a set of strict design requirements that all have to be met together. These generally conflict with one another. There are therefore trade-offs intrinsic to every solution design, which set configuration-specific limits on the performance potential. The configuration-dependent limits of any specific technological solution—especially complex solutions—are therefore usually much lower than any absolute limits.

Upon its introduction, a newly invented solution is likely to be far from its configuration-dependent limits. Analysts speak in terms of the 'technological distance' from limits when explaining the pace at which technological improvement occurs, the rate of return on investment and the willingness to invest in performance improvement. All of these are greater the further the technology is from its limits.[3] An illustration is provided by the contrasting situations of some familiar technologies. Reversible heat engines operating between two tempera-

---

3   Technological distance has therefore been invoked as a factor in explaining shifts in investment between projects and even between different sectors of the economy.

tures are close to their configuration-dependent limits, which are now well understood. The scope for further improving the energy efficiency of reversible heat engines is all but exhausted.[4] In contrast, the computer chip was far from any limits at the time of its introduction[5] and appears still to be so today. Indeed, it is so far away that it is not yet clear what the limiting factors will be.

An important lesson is that the nature of technological limits often only become clear as these are approached, at which point designers start to recognise how conflicts and interactions that are intrinsic to the initial solution design narrow down, and ultimately close off, the set of improvement options. Moreover, an improving understanding of the nature of the limits faced is important for the specification of the technological challenge ahead. Another important aspect of the specification is to be clear about the precise needs that the new technological solution should meet. As a generalisation, if the challenge to technological innovation can be specified precisely and articulated clearly, the chances of focusing R&D efforts where they are most likely to yield new breakthrough solutions are greater and so, too, is the likelihood that a new solution will emerge quickly. Gilfillan made this point as long ago as 1937 using the example of aircraft navigation (Ayres 1998b), at the same time demonstrating that it is possible to use indicators, such as the quality of the need specification, as pointers to technological futures. Once aeroplanes had been developed, there was a clear need for a technology that would enable an aircraft to land safely in conditions of poor visibility, such as at night or in fog. It was also possible for Gilfillan to identify a set of possible approaches for finding a solution the implementation of which would not violate any known physical law. It does not matter that his initial list of sketched solutions to this particular technological challenge did not include the one that was ultimately successful. Gilfillan's general point was that such a technology was needed, that there were no theoretical barriers and, therefore, that one could expect a solution to be developed within a few years (Ayres 1994). In fact, the time taken between the articulation of this particular technological challenge and its solution was a matter of a few years only.

All technological solutions therefore have an innovation 'life-cycle'. They are 'born' as a radical new solution that offers possibilities for performing a new task or for better performing an old task. They are then improved, slowly at first, but on an accelerating pathway as experience with the new technology builds with

4   The theoretical (Carnot) limit for any reversible heat engine operating between two temperatures is given by the difference between the temperatures, divided by the absolute temperature of the hot reservoir.

5   The distance was so great and the improvement potential so large that, in 1964, the Director of Research at Fairchild, Gordon Moore, predicted that processing power would double and production cost would halve with each doubling of cumulative output until at least the 1980s. This became known as Moore's Law. The prediction was borne out by the history of chip development. By the early 1980s, there had been 11 such doublings of cumulative output.

growth in cumulative sales and through the feedback mechanisms just described. The technology goes through a stage of rapid development (adolescence) to reach maturity (adulthood). Throughout this period, investment in further improvement is worthwhile since the benefits exceed the costs. Ultimately, however, the improvement rate will begin to slow down as the market for the solution becomes saturated and the technology nears the limits of its optimisation potential. The technology is now fully mature. Diminishing returns are experienced from further investment in improvement. At this stage, the past success of the technology may also undermine it. The scale of existing use of the technology will exaggerate any negative externalities, such as resource requirements or pollution implications. In the end, competition with a newly introduced radical solution will finally undermine the old solution. Most mature technologies shrink back into a niche or die.

As the old technology reaches maturity, then pressure builds to find a radical new solution. When one is found and introduced, the former solution becomes 'obsolete' and its role in the economy usually dwindles or consolidates in narrow niche applications. In the wider context, however, the new technology gradually takes over. A new phase of incremental innovation begins focused on optimising the new solution and the cycle starts over. For illustrative purposes, Figure 3.1 describes the historical innovation process in respect to illumination, the task of

Figure 3.1 **The lighting S-curve**
Source: Betz 1993

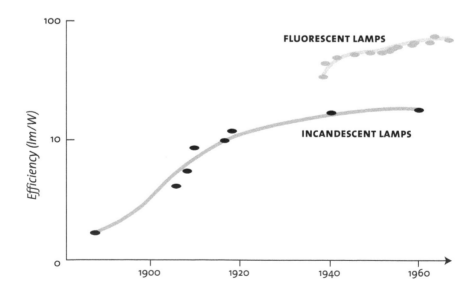

providing light. Incandescent lamps were introduced in the 1920s.[6] Their performance—in terms of energy efficiency and cost—was rapidly improved in the early years, but improvement soon began to level off. The radical new solution was the fluorescent lamp. In the early years, fluorescent lamps had many teething problems, especially as regards reliability. But their energy efficiency soon made them the most cost-effective means of providing high-intensity lighting. Incandescent lamps still occupy a role in home lighting, because they offer a gentler light that is not so harsh as that from fluorescent lamps. But they now occupy only one niche in the overall market (Fussler with James 1996).

DIFFUSION

Technology diffusion refers to the process and pattern of the adoption and spread of a technology over time, in space and across different user groups. It comes in two forms. Pure diffusion refers to a solution that creates an entirely new market. Replacement diffusion refers to a solution that enters an existing market, which it may expand but in which it also competes with established solutions for a share of the total market. No innovation is taken up instantaneously. Rather, it spreads gradually over space and time at rates that are influenced by many factors relating to the characteristics of the technology (such as its costs and benefits, compatibility with existing contexts, complexity, appropriability) and whether it is building new markets or competing in old markets to replace established technologies. In cases where technology is of any economic or social significance, diffusion often takes decades. For large-scale infrastructures, such as road, rail and pipeline networks, diffusion may take 100 years or more. Typically, diffusion takes longest but extends furthest in the 'innovation centre' where the initial innovation occurs, from which it spreads, often via a hierarchy of 'sub-centres', to the 'periphery'. In the periphery, a catching-up process begins. This generally sees a faster rate of adoption but a lower adoption peak (Hägerstrand 1967).

When a technology is first introduced, the rate of adoption by potential users is typically low. The new technology is unknown, untried and unproven. The first units will be costly to produce, so the selling price will be high compared with subsequent units and older technologies. However, the new technology is likely to offer advantages to some potential users and will usually appeal to a minority that is less price-sensitive and less risk-averse than the majority. If enough of these early adopters buy the new technology, producers will start to move along the technol-

6 Incidentally, this development was dependent on earlier breakthroughs in the field of electricity and, perhaps less obviously, pyrotechnology, since it was only possible to make use of metals such as tungsten (used for making the filament in light bulbs) after achieving the capacity to generate and control temperatures of 3,000°C.

ogy learning curve. Through innovation and marketing, sales will slowly increase. If sales increase sufficiently, the synergies between adoption and innovation can become self-sustaining. A take-off point is reached after which the adoption rate tends to accelerate away. In the period immediately after take-off, rapid increase in the number of adopters can be sustained by innovation made possible through scale economies and R&D. To take full advantage of the growth potential and to sustain it, producers may also expand production capacity ahead of demand. Intense competition among producers and the fight for market share will drive innovation and force prices down. After a time, however, the pool of potential adopters left to take up the technology will begin to dry up. The technology will also be closing in on its technical limits. The adoption rate no longer accelerates but slows down. The virtuous cycle that has sustained the diffusion loses momentum.[7] Further investment in the technology tapers off as the adoption curve plateaus out. For virtually all technologies a plot of cumulative sales against time therefore yields an S-shaped adoption curve, which mirrors that of the innovation curve.[8]

Empirical studies of technology diffusion reveal several consistent findings. There is a very large number of technological 'non-starters' that never succeed in building a market or penetrating an existing market (Grübler 1998). Moreover, the period between invention and widespread use, the first plateau of the S-curve, is very long (Fussler with James 1996). This is true both for pure diffusion processes and for replacement diffusion. Marchetti and Nakicenovic (1979) have proposed a generalised model of multiple competing technologies based on consistent findings about the time taken for new technologies to expand from a 1% share of the overall market to a 50% share. In the case of road and rail transportation systems, these periods are of the order of 55–65 years. This finding is important. It points to the long lead-times needed when preparing for technological change, which are significant when wanting to use technology to reduce the environmental burden. Only if a technology is successful in taking market share away from eco-inefficient solutions, which it both displaces and replaces, will it contribute to environmental protection (Box 3.1). Technology diffusion is therefore of great importance for sustainability, alongside innovation. Since long periods are involved both in developing new solutions and in building markets for them, it is important to plan well ahead, to work simultaneously on innovation and adoption challenges and to integrate adoption criteria into design processes.[9]

7   Competition among producers for the remaining market, made up of late adopters and replacement demand, tends to intensify as the technology comes to the end of its diffusion. A round of industry consolidation and rationalisation is likely to be necessary in order to maintain industry profitability.

8   In effect, the innovation and adoption processes are mutually supporting. Together they describe a general pattern of technological development that repeats across technologies, sectors and settings.

9   Empirical studies that have charted the history of significant technological events in the development of specific sectors or markets have also found that there is often a long

IN REGARD TO THE SUCCESS AND FAILURE OF TECHNOLOGIES, THE COMMENT HAS been made that the history of technological change is systematically biased in favour of technologies that have been successful. These are the ones whose wide-scale production and use is recorded in physical artefacts and whose impacts have been great enough to warrant attention. This can give the false impression that most technological solutions are successful. Nothing could be further from the truth. Many start the technology race, but few finish it. Unsuccessful technologies are quickly forgotten and leave few traces.

This point is worth developing further, since there are other serious misperceptions regarding technological change processes. Because of the competitive mechanism inherent in a capitalistic economic system, there is a continuous pressure to search for improvement in the productivity of all priced factors of production. Since producers face costs in using energy and material inputs—albeit not always full costs—the historical record shows a trend of steady efficiency improvement in the use of natural resources over a fairly long period. On the basis of these long-term average trends, economic forecasters often assume that the average past rate of progress will continue into the future. This assumption is built into forecasting models. This means that, when making forecasts of, for example, future energy use, any expected annual growth in demand for energy services is immediately offset by the average past rate of efficiency improvement, which is anticipated will continue to apply.

This is misleading in several ways. Long-term averages give an impression that improvement is achieved smoothly and at an even rate. In practice, progress is typically episodic with cycles of fast and slow progress. Progress may also be harder to achieve in the future as past progress has taken us closer to configuration-dependent limits. Equally, the assumption of an autonomous technological improvement gives the impression that innovation is effortless, costless and without risk. Grübler comments that this is akin to assuming that technology is a 'free good' that drops like 'manna from heaven' (Grübler 1998). Again, nothing could be further from the truth. Innovation demands enormous effort, is expensive and is fraught with risk.

Box 3.1 **Misperceptions about technological change**

TECHNOLOGICAL CHANGE AND DEVELOPMENT

It is useful to study individual technologies. Nonetheless, no individual technology or innovation alone—however important—is capable of having much impact on the pathway of social, economic and environmental development. One of the many seminal contributions of Schumpeter was to draw attention to the importance of synergistic clusters of related innovations whose combined impact on development processes are greater than the sum of their constituent parts. Schumpeter drew attention not only to technological innovations, but also to related institutional,

(although variable) time-lag between invention and first commercial introduction (Rosegger 1996), which adds to the lag involved in building markets and replacing older solutions. All of these lags imply that a long lead-time is needed to change the environmental burden of economic activities through technological means.

organisational and managerial innovations in cross-enhancing and mutually reinforcing combinations.[10] Moreover, he used these to explore the links between innovation processes and the repeated, fast–slow pattern of historical development. He considered both to be outcomes of processes within capitalism and therefore both also to be intrinsic outcomes of capitalistic economic systems.[11] Schumpeter described socioeconomic development as an uneven process that follows a repeating pattern. A phase of rapid growth accompanied by a restructuring of the economy is followed by a period of slower growth and later by stagnation and even recession.[12] Schumpeter linked the periods of rapid growth and restructuring to the introduction of clusters of synergistic innovations, which were, in their turn, responses to the redirection of creative efforts away from old solutions where diminishing returns had already set in and toward the discovery of new solutions and the creation of new markets (Schumpeter 1935, 1939, 1942).

LESSONS FROM THE PAST RECORD OF TECHNOLOGICAL CHANGE

For the purposes of this chapter, it is not worthwhile going into the details of the historical relationship between technological innovation and the patterns and pathways of economic development processes. Excellent summaries are available elsewhere (e.g. Grübler 1998). It is sufficient for our purposes just to highlight the key themes that have emerged from studies of the relationship. First among these is that the history of development is marked by a succession of phases, each characterised by its own distinctive cluster of synergistic technological, institutional and organisational innovations, which co-evolve throughout that phase and

10   Building on this concept, Freeman and Perez propose a hierarchical taxonomy of technological innovations related to a hierarchical set of economic and social impacts. At the higher levels, they recognise changes in 'technology systems' involving clusters of technological, managerial and organisational innovations. These can affect developments in entire economic sectors. However, it takes several new technology systems in synergistic combination to impact on economic and social development as a whole (Freeman and Perez 1988).

11   Competition and the search for higher returns within a competitive capitalist economic system force innovation. Innovation creates differential productivity increases that create an economic disequilibrium. Schumpeter wrote in terms of a process of 'creative destruction' as old technologies and their infrastructures that offer lower returns are replaced by newer and better technologies and investment shifts to markets and sectors that are the fastest growing and offer greatest returns. Later, Fourastié proposed a mechanism by which differential productivity improvement could lead to restructuring across sectors through the diversion of productivity gains to new sectors (Fourastié 1949).

12   The pattern had already been observed by Kondratiev who saw economic growth occurring in waves and had proposed a long-wave theory of economic development (Kondratiev 1926).

give rise to a coherent, functionally consistent, socioeconomic development system and trajectory. There is a definite path-dependence in development processes (Nelson and Winter 1977). Although each phase represents a new approach, each also develops from and builds on earlier developments. Past solutions and decisions influence new trajectories. Each new phase of development therefore evolves out of former phases of development, which influence it. In turn, it will influence future phases. Each phase is characterised by clusters of dominant technologies—sets of interrelated technological, infrastructural and organisational innovations—that drive output and productivity growth throughout the period. But within each phase there are vestiges of earlier technology vintages and, also, harbingers of technologies to come, often being worked out and tested in niche markets protected against competition from the dominant technologies of the day. Empirical studies show that, overall, the temporal envelope of any particular technology cluster spans up to a century with its main growth period covering about five decades (Grübler 1998). In his recent book, Grübler describes a succession of four such clusters in the post-industrial era and conjectures on the possible emergence of a fifth.

The historically observed pattern of successive phases of development begs some questions about the possibility of achieving a technological transformation that would reduce the claim on eco-capacity. Especially, it forces us to consider our current position in the relevant cycles and the development level that we have already reached. Within any particular development phase, technological change is in harmony with the prevailing techno-economic paradigm. Technology designs out of step with that paradigm will not be able to compete. Only when the scope for further returns to the existing regime start to run out or when problems emerge that are intrinsic to it will attention turn to constructing another regime that overcomes the limitations of the last. In principle, then, it will be easier to introduce new technology clusters when an established phase of development is already coming to its end. This is likely to be signalled by a period of 'innovation lethargy'. Many analysts consider that Western societies and economies are approaching the end of the current phase of development (Fussler with James 1996). However, another factor that might affect the chances for new technology clusters is the systemic complexity of the established regime of accumulation. Existing technologies have co-evolved with each other and the prevailing context. This creates systemic harmony, but also systemic interconnectedness. Whereas harmony is useful when the dominant regime is far from its limits, interconnectedness is potentially an impediment to change when the need is to shift to another regime (Frankel 1955). The more developed an economy already is, the more difficult it is likely to be to introduce new technology clusters and to switch to a new regime and the more pain is likely to be felt during subsequent restructuring.

Nonetheless, the past record of technological change is nothing if not one of successive waves of technological restructuring. In spite of all the inertias, history shows that there is a finite life-span for each technology and for each technology cluster. Ultimately, diminishing returns set in to every techno-economic paradigm and change is inevitable. Technological change and restructuring are intrinsic to the mechanism of 'creative destruction' that is inherent to capitalistic economic systems (Schumpeter 1942). The lesson of history is that, given the right incentives and facilitating conditions, pervasive technological transformations can be implemented within one or two human generations (Grübler 1998). The issue, therefore, is not so much one of needing to initiate technological and structural change, but rather one of influencing change that will happen anyway and directing it toward desirable societal objectives, at the same time ensuring that change in this broad direction proceeds at a sufficiently rapid pace. An important factor is that the critical incentives and facilitating conditions for technological change are amenable to policy influence. They are socially constructed and, in principle, can be reconstructed.

TECHNOLOGICAL CHANGE AND THE ENVIRONMENT

However, there is no guarantee that the next wave of technologies will be one that helps to reduce the environmental burden. The history of technological change shows the relationship between technology and the environment to be complex and paradoxical (Gray 1989). It is a two-way relationship. Technologies use resources and impose environmental stress. But technology is also a response to resource constraints and environmental problems. Technological change has been grouped into four categories: those that augment resources, those that enhance productivities, those that diversify products and production and those that (directly and indirectly) enlarge markets (Grübler 1998). It is by no means clear or automatic that all technological change benefits the environment. Historically, technological change has improved resource productivities and opened up new resources. It has repeatedly freed society from immediately pressing environmental constraints, enhanced productivity and paved the way for spectacular increases in human numbers and economic activity. It would be untrue to say, though, that environmental constraints have been permanently removed. Rather, technological change has shifted the resource base of production, which has relaxed (perhaps only temporarily) some constraints while simultaneously making other, less immediate constraints more important. By enlarging markets and changing the scale, structure and spatial pattern of economic activities, technological change has, indirectly, also added to the overall claim on eco-capacity and increased society's longer-term vulnerability to new constraints.

This has implications for our understanding of biophysical limits and their interrelation with other forms of limits, including technological limits and institutional limits. In their relationship with the environment, humans are unique among species. Technology enables us to modify the environment to fit our needs and to shift the resource basis of production to overcome biophysical constraints. Humans not only have 'needs' but also 'wants', which vary across individuals and change over time. Since human needs and wants are influenced by what is possible and affordable, they co-evolve with the development of technology and the level of affluence. Limits are therefore not imposed by biophysical factors alone but rather by a complex and dynamic interplay involving environmental capacities, technological capacities, institutional capacities and culture.[13] There are likely to be few immutable limits on human activities. Nonetheless, all human activities are embedded in contexts and, so, context-specific limits are relevant. The message of limits is not that these are immutable but that at any given time societies are limited in how they can deal with them. In effect, the importance of limits is as a source of working hypotheses about barriers to progress and opportunities for overcoming these (Weterings and Opschoor 1992).

The fact that improvements in the eco-efficiency of individual technologies will not automatically translate into lower overall claims on eco-capacity also has implications for the concept of 'sustainable technology'. The concept can never be absolute since, as well as the attributes of the technology, the environmental stress it imposes depends on the applications context. This includes the adequacy of the social support systems and institutions where it is applied and the scale of use. A technology that is sound when used on a small scale or for a limited period in one place may be unsustainable if used on a large scale, over a long period, or in a different place. Thus, one of the primary functions of technology assessment is to estimate in advance the limits on the scale of use that are compatible with sustainability and to modify the subsequent evolution of the technology so as to moderate the growth of its impact with scale of use (Brooks 1992). This stresses that technology diffusion and use are as important as innovation and design issues. The management of all of these will become increasingly important in societies' attempts to achieve sustainability. It also implies that technology cannot be *the* solution, of itself. Complementary structural and cultural changes will be needed to integrate sustainability values into the economy, to build markets for eco-efficient products and to influence behaviours and lifestyles so that gains in resource productivity can be captured.

---

13   Returning to the discussion in Chapter 2 on the need for operational indicators for sustainable development, it is clear that the search is complicated by the enormous number of possible social, economic, technical and cultural variables involved, their interactions and their interdependencies (Brooks 1992).

SOURCES OF
TECHNOLOGICAL CHANGE

From the perspective of designing a programme aimed at influencing technological change processes, it is interesting also to look to the historical record for evidence of the past sources of technological change. Some of these have already been mentioned, such as the level of investment in basic and applied research and the socioeconomic context. An observation with important implications is that technology development is an inefficient process. It is also one that is difficult if not impossible for central authorities to direct successfully. First, technology is not free. Technology must be deliberately induced at cost to the innovators. The costs are high because of the high failure rate of new innovations. Second, history points to a very low success rate of centralised efforts in regard to technology planning. There is substantial evidence of earlier, well-intended but unsuccessful attempts at technology planning by centralised authorities, including attempts to pick and back 'technology winners'. However, the record shows that, because uncertainty and experimentation are inherent to the process, successful technology development depends heavily on decentralised decision-making structures and processes of information exchange. Markets and informal information networks among technology suppliers and users are important to the processes of weeding out inferior technologies, selecting superior alternatives and learning how to improve these.

Studies of technological change therefore repeatedly stress the importance of the firm as the key organisational entity in carrying innovation forward. Nonetheless, it has been found that radical solutions rarely arise from firms with existing interests in the same market. Radical innovations are, instead, most often introduced by firms new to the market or by new constellations of firms. Moreover, while firms are the key organisational entities, the process of developing new solutions involves interactions between demand and supply, technology producers and users, private and public R&D and knowledge and competencies internal and external to the firm (Freeman 1994). Interactions are orchestrated via 'social networks' involving many diverse actors who may be either proponents or opponents of solutions. An important phenomenon is that, over time and through communication among its members, a network develops a 'joint-technological expectation' (Rosenburg 1982). This influences and is influenced by commonly shared understanding, visions and missions. Studies of past technological change show that these expectations can become self-fulfilling.[14]

---

14   Everyone within the network expects technology to develop in a certain direction and, so, focuses efforts and resources in that direction and makes plans in the expectation of realising a solution.

TECHNOLOGICAL CHANGE
AND COMPETITIVENESS

It is also useful to draw attention to the relationship between technology development and competitiveness. The relationship is important at several levels of
organisation, but especially for the enterprise, the nation-state and economic
blocs. From the discussion so far, it is clear that innovation efforts can be focused
either on improving existing technologies or on developing radical new technological solutions. An important finding from empirical studies is that the most
successful enterprises are ones that run on parallel tracks, going forward with
incremental change in the short run, but also investing in long-term, strategic
planning and R&D for breakthrough technologies (Morone 1993). This long-term
view is necessary because it takes decades for technology to travel from initial idea
to commercially viable product. Another lesson is that early innovators typically
dominate their industry over long periods. By way of example, the origin of the
disproportionate share of world markets for chemical and pharmaceutical products
held today by Swiss and German companies dates to leadership from the 19th
century. The US has similarly held on to its pioneering advantage in agricultural
machinery, long-haul civil aircraft and computer software (Ayres 1998b).[15] The key
to a pioneer's advantage lies not in having 'good' technology, but 'better' technology and always keeping ahead in the technology race by making use of the learning
opportunities of leadership to keep ahead.[16]

THE CASE FOR PUBLIC SUPPORT FOR
SUSTAINABLE TECHNOLOGY DEVELOPMENT

Past government programmes aimed at inducing technological change have not
been successful. Yet it is also clear that sustainable technologies are needed, that
current rates of progress in developing them are too slow and that trends in
demographics and consumer aspirations are on a collision course with biophysical

15   In cases where leadership has been lost, this is often due to the development of a
      competing new technology, such as when the US and the UK gained a significant share
      in chemicals by pioneering synthetic fibres or when Japanese companies used large-scale
      automated semiconductor manufacturing to overcome US first-mover advantage in
      sound recording (Ayres 1998b).
16   Rather than using a leadership position to maximise short-term profit, companies can
      use it to stay ahead of competitors by sharing the benefits of their learning—in the form
      of lower costs and better product performance—with customers. This has the effect of
      holding the firm's market share in the short term and maximising further learning
      possibilities, a combination that allows leaders to move fast enough along the learning
      curve to stay consistently ahead of competitors over long periods.

limits (Fussler with James 1996). Most existing technologies were never designed with environmental compatibility in mind and it was never envisaged that they would be so generally used. Many of today's important technologies are already mature. But little progress is being made in finding environmentally compatible replacements to fulfil the same functions. Sustainable growth is possible so long as technological progress can be made to result in a faster dematerialisation of GNP than overall growth of GNP and so long as eco-capacity constraints are not overreached. However, progress in dematerialisation today—even in the most advanced economies—lags behind economic growth. A large part of the reason is that incentives are distorted, operational control is fragmented and short-termism is institutionalised by today's arrangements. The question arises: How can sustainable technologies be induced when innovation is still focused on incremental improvement? However difficult the task of trying to shift the balance of R&D efforts in favour of radical innovation and breakthrough technologies, this is one case where 'doing nothing' is probably not an option.

But to what extent is 'shifting the balance' a public responsibility? Public support for R&D is justified—indeed is necessary—when the major beneficiary from new technology is the public and when private firms with the competence to undertake the development work are likely to under-invest relative to the socially efficient level of investment (Ayres 1994). A criterion for public support, therefore, is that the social benefits should be significant in relation both to the costs of support and the privately appropriable benefits. Private firms are likely to under-invest in the face of high market uncertainty or if the benefits of their R&D investments are difficult to appropriate or to protect against 'free-riders'.[17] They are likely to under-invest, also, in the case of infrastructural projects involving substantial network externalities. All of these conditions apply to sustainable technologies. Public investment in R&D is also justified because society must consider the long-term future more than firms must, and because society values non-economic spin-offs from long-term R&D more than firms do. Knowledge is one such spin-off and it has special attributes: knowledge is a pure public good, there are increasing returns to knowledge and knowledge is non-rival.[18] This is the basis for many analysts' belief that substantial government involvement is needed if sustainable technologies are to be developed within desirable time-frames, and is justified. The question remains, however: How should a public programme be designed to be

17 Government itself is implicated in creating these unfavourable conditions. For example, past and present government policies have exaggerated market failures rather than correcting these. Future policy is a major influence on the future market for sustainable products and services, but is uncertain.

18 Unlike other resources, knowledge is not diminished by use. The case for investment in R&D generally has been made by several analysts, perhaps most notably by Brooks (Brooks 1992). Brooks's arguments rest especially on knowledge as the key both because it can substitute for material resources and assimilative capacities and because it is a resource that is not depleted by more intensive use.

most effective? To answer this question we need to draw on the lessons from studies of past technological change.

LESSONS FOR
PROGRAMME DESIGN

What do the lessons we have learned so far suggest for the design of a programme aimed at inducing sustainable technologies and at evaluating the contribution that technology might make to sustainable development? Perhaps the broadest implication is that the role of government should be to influence technological and development trajectories indirectly, by encouraging innovation in respect to the innovation process itself, rather than by picking and backing prospective technological winners.[19] The role should be to try to shift the balance of the R&D efforts of firms and technology developers toward longer-term issues, focus their attention on the long-term challenge of sustainability and increase the chances of success with their innovation efforts, especially by facilitating the networking process. Decision-making should be left to network members and to firms undertaking R&D. Any public support should be provided in kind rather than as financial backing and should be conditional on private actors committing their own resources. In effect, firms and networks should be the key organisational entities and decisions should be taken in decentralised decision-making structures.

The activities undertaken within a public policy programme, therefore, should relate to this general facilitating role:

▷ Helping to build new networks that bring firms, research institutes, public agents and users of technology together

▷ Helping network members to reach common problem definitions, shared understandings of innovation challenges, shared visions of possible futures and shared expectations

▷ Providing information, tools and approaches that network members can use to explore possible futures and to translate long-term visions into short-term action steps

▷ Encouraging innovators to integrate technological, cultural and structural aspects, so that these develop in the same direction and to a consistent time-frame

19 Supporting the innovation process is especially important given the difficulty, perhaps impossibility, of selecting technological winners in advance. While example technologies are needed to pursue this objective, and it is only sensible to use promising technologies as examples, the main goal of public policy support should be to improve the capacities for sustainable technology innovation and not to advance one or two selected technologies.

▷ Helping to create conditions conducive for technologies emerging from R&D activities by liaising with policy-makers about implementation needs and sustainability benefits

In addition, a public policy programme should be constructed as a 'learning-by-doing' activity, with a charter to evaluate its own performance, learn from mistakes and disseminate lessons learned so that these can be used elsewhere. This is necessary in order to leverage its effectiveness.

In order to have best chance of being effective in achieving its goals, a programme should:

▷ Have a broad (multi-sector, multi-resource and multi-need) scope, sufficient to explore synergies between technologies and technology clusters

▷ Take a long-term perspective, consistent with the 50-year period needed for radical breakthrough innovations

▷ Set ambitious targets as 'working hypotheses' of the technological challenge inherent in sustainability

▷ Focus on major sources of unsustainability in areas such as nutrition, energy and materials

▷ Ignore existing technological solutions and, instead, develop new solutions from new starting points, such as the 'need' to be met or the 'task' to be performed

▷ Emphasise diversity in exploring solutions

▷ Emphasise 'learning-by-doing'

▷ Use models from the natural world as a source of inspiration, since nature represents an amalgam of sustainable solutions.

Chapter 4

The STD programme design and approach

The outcome of the background studies was a decision by a consortium of Dutch ministries to proceed with a full 'Sustainable Technology Development' programme. Sustainable development had been adopted as a policy goal of the Dutch government in 1989 and the role technology could play in achieving environmental protection had been stressed in technology and environmental policy since then (NEPP 1989; Tweede Kamer 1991). Nonetheless, the background studies for the STD programme revealed an important gap in policy efforts. Policies in place by the early 1990s were aimed at improving the environmental performance of existing technologies over two different time-horizons. They promoted innovation on a short-term 'track', aimed at end-of-pipe clean-up, and a medium-term 'track', aimed at process-integrated improvement. Both tracks are essential to environmental protection efforts. But, even at best, improvement of existing technologies offers only temporary respite from resource and environmental constraints. Efforts along these tracks might win time, but sustainability ultimately depends on path-breaking innovations, which provide longer-term solutions that are inherently better than existing ones. Such radical innovations have long gestation periods and developing them calls for sustained efforts. The background studies clearly demonstrated the need for a 'third' innovation track, revealing this to be a crucial policy gap that should be filled. Policies to promote technological innovation and sustainable development can be fully integrated only through support for innovation on this third track, since only if sustainability concerns are integrated into technology designs from the outset can conflicts and trade-offs between economic, social and environmental goals be minimised.

THREE TRACKS FOR INNOVATION

The three innovation tracks have different time-horizons and objectives. The short-term track is mostly concerned with issues of care and good housekeeping. In industry, this corresponds to the fine-tuning of operations through mechanisms such as quality management, maintenance, auditing and efficiency drives, which typically have a time-horizon of up to five years. The short-term track can deliver limited environmental protection through 'end-of-pipe' measures. The medium-term track is concerned with process- or product-integrated technological improvement and reorganisation. Nonetheless, it is confined to improvement within the framework of existing infrastructures and technologies. The typical time-scale is five to twenty years. Medium-term innovation can reduce environmental stress by delivering 'environmental technologies'. The third, long-term, track is concerned with fundamental renewal of technologies and organisational arrangements. By breaking with past practice, it can deliver leaps in performance. Innovation for technology renewal involves redefining existing technology development trajectories and provoking new ones. The time-scale is 20 years or more. In principle, renewal can deliver 'sustainable technologies': ones designed from the outset to be intrinsically compatible with eco-capacity constraints.

Because the three tracks have different time-horizons and objectives, they represent complementary, not alternative, innovation fronts. Improving the performance of existing solutions is critical for holding back environmental change, which wins time for developing long-term solutions. There is therefore no inconsistency in providing policy support for all three tracks simultaneously. Moreover, the long lead-time for developing sustainable technologies means that there is no time to lose in beginning work on technology renewal: the more so because the natural urgency of innovation efforts on the long-term track tends to be lower than for other tracks, since competitive and regulatory pressures for efficiency and performance improvement are weaker. Figures 4.1–4.3 represent the different time-horizons and objectives of the three tracks and demonstrate their interdependence and compatibility. Figure 4.1 illustrates this schematically in the case of emission reduction, showing that each innovation track can contribute to reducing emissions, but within a different time-horizon. It also illustrates that there are limits to the emission reductions that are achievable via end-of-pipe and process-integrated responses. Figure 4.2 suggests limits on the eco-efficiency improvement potential of current technologies and sets targets for future technologies. The STD programme expects that fundamentally new systems solutions will be needed to deliver eco-efficiency improvements over today's solutions of a factor 4 or more. If sizeable leaps in performance are to be made in the first half of the 21st century, the build-up for these must begin now. While improvements on

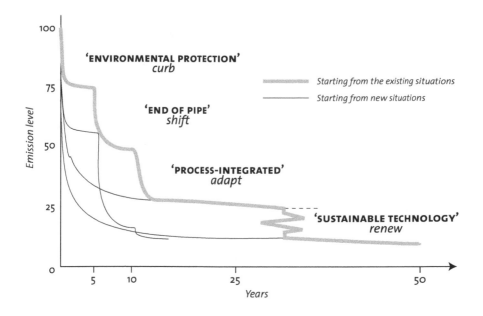

Figure 4.1 **Emission reduction with technology**

Figure 4.2 **Technology for sustainable development: targets**

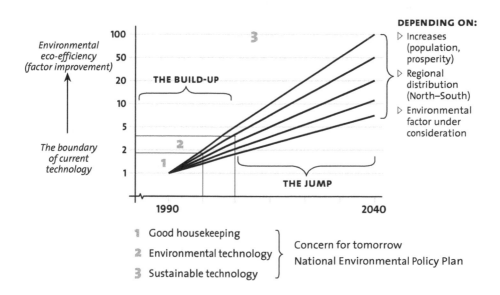

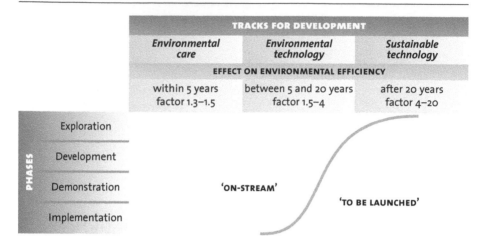

Figure 4.3 **Development tracks for technology for sustainability**

the first two tracks are already 'on-stream', the bulk of the effort on the third track is still 'to be launched' (Figure 4.3).

The three tracks call for entirely different approaches to innovation: different intensities of effort, different organisational arrangements and different modes of working. They call for different actors to be involved in innovation and for actors to play different roles. The characteristics of each of the different innovation tracks are summarised in Table 4.1. Innovation for technology renewal implies a search for systems solutions consistent with a shift in the development paradigm.[1] Successful development of eco-efficient technologies therefore depends on pre-empting future structural and cultural conditions and integrating these into designs, since these will determine the possibility, desirability and acceptability of the innovation (Schwarz and Thomson 1990; Schwarz 1992). The challenge for innovators concerned for technology renewal is therefore of a more conceptual nature than that of making incremental changes to technologies within known economic, socio-cultural and infrastructural frameworks. It requires thought and action at a strategic, 'systems' level and at intersectoral and macro scales rather than at the usual intrasectoral or micro scales. The driving forces for innovation are different, too. Those for technology renewal are less tangible and less immediate than the drivers of short- and medium-term innovation, such as the pressure

1 A sustainable development paradigm will be rooted in fundamental shifts in the relative cost of dominant production factors and in societal values, attitudes and behaviours in respect to the environment. The acceptability of eco-efficient technologies is directly connected to structural and cultural conditions, which are dynamic.

Innovation trajectory / Characteristic	CARE/ OPTIMISATION	PROCESS AND PRODUCT IMPROVEMENT	RENEWAL
TIME-HORIZON	0–5 years	5–20 years	20–50 years
CHARACTER OF INNOVATION	End-of-pipe	Integrated in process/product	
MEANS OF IMPROVEMENT	Application	Adaptation and improvement	New systems and paradigm shifts
BURDEN REDUCTION POTENTIAL	Factor 1.5 (30%)	Factor 5 (80%)	Factor 10–50 (90%–98%)
NATURE OF CHALLENGE	Operational	Implementational	Conceptual
ORGANISATIONAL LEVEL	Plant/line managers	Divisional managers	Board/CEO
DRIVING FORCES	Cost reduction, performance improvement, public image	Belief in long-term developments and shared prospects	
ADMINISTRATIVE INCENTIVES	Direct regulation	Regulation, taxation, sectoral (voluntary) agreements, policy declarations	Strategic, co-operation agreements
INNOVATION NETWORKS	In-house	Internal/external communication	Joint working and co-operation
DEGREE OF UNCERTAINTY/RISK	Low	Low to moderate	High
SCOPE OF ACTION	Micro	Micro/macro and intra-sectoral	Macro, inter-sectoral
SPECIFICATION LEVEL	Process	Process, product, material	Function, need
ACTORS IN TECHNOLOGY DEVELOPMENT	Private enterprise	Private enterprise, technology institutes, universities	Universities, institutes of technology, private enterprise

Table 4.1 **Characteristics of the different innovation tracks**

to reduce costs, become more efficient or meet regulatory environmental standards. The drivers of technology renewal are related mostly to beliefs about long-term trends and shared development prospects. There is also much greater uncertainty and risk in respect to long-term innovation, because of the greater number of uncontrollable and intangible factors.

For these reasons, long-term innovation is likely to be taken up only if those at the very highest levels in companies recognise it to be important. Whereas innovation on the first and second tracks is undertaken largely by private enterprise acting alone, innovation on the third track is more likely to be undertaken by business consortia, involving several business enterprises working together at the pre-competitive level, together with universities, institutes of technology and government bodies. Co-operation agreements among innovators can help to reduce uncertainty and lower the cost involved in long-term innovation. Government support to set up such co-operations is therefore an appropriate means of administrative intervention. Government, which in principle has the major responsibility for safeguarding society's long-term interests, can also play a role in drawing business leaders' attention to sustainability concerns. One way of doing this is by providing an outline set of conditions for long-term sustainability and initiating a first set of strategic reviews of how critical needs in key areas might be met in the future.[2] Through such reviews and the lessons they provide about how to set up a review process, government may stimulate a ripple-in-the-pond effect in favour of wider technology renewal. This was the approach taken in the STD programme.

PROGRAMME AIMS, OBJECTIVES AND ADMINISTRATIVE ARRANGEMENTS

Differences in the characteristics of innovation tracks justify different organisational, managerial and operational approaches to providing public support for innovation. A public programme of support for long-term innovation should have critical mass, wide-ranging scope, a long-term time-horizon and it should be based on challenging eco-efficiency targets. It should be process-oriented.[3] In this regard, three aspects are especially important. The innovation process should encourage a co-evolutionary 'systems' view of the dynamic relationships between technology, culture and structure. It should engage not only stakeholders who will be involved in the technology development, but also those who will be affected, directly or

---

2   Provocative assumptions about possible breaks in current trends can be used as the basis for such reviews: for example, the assumption that in the long term future fossil energy will no longer be available to the economy or that it will no longer be permissible to dissipate toxic materials into the environment.

3   While it must influence mind-sets and actions in today's world, a programme could not be expected to bring about significant reductions in the levels of environmental stress until sustainable technologies facilitated, directly or indirectly, by programme activities have achieved significant market penetration. A programme should therefore be focused on influencing the micro-innovation context and innovation methods of enterprises and not on promoting specific technologies.

indirectly, by the technology.[4] And it should encourage a translation of long-term solution concepts into near-term action steps toward realising them.[5] However, decision-making should be decentralised. Firms and enterprises should be the 'innovation champions', not the government.[6] Since, any programme would be breaking new ground, it should be conceived as an exercise in learning-by-doing and be charged with developing, testing and refining new innovation methods.

These defining aspects are reflected in the STD programme aims. The STD programme was established as a research programme with a mandate for 'learning-by-doing'. Its aims were to:

▷ Integrate sustainability policy and technology policy

▷ Develop, demonstrate and evaluate methodologies for influencing innovation processes in the direction of sustainable technologies

▷ Initiate new technological trajectories related to key areas of need

▷ Engage and involve stakeholders in innovation processes

▷ Demonstrate that sustainable technologies are possible, in principle, if research, development and innovation processes are appropriately oriented and resourced

▷ Disseminate and communicate programme results, nationally and internationally, among innovators, policy-makers, opinion leaders and partners in implementing sustainability

On the 'doing' front, the aims of the STD programme were limited to enabling technology developers and other stakeholders to gain their first experiences with the search process and to making some first strategic reviews of a few key need fields. On the 'learning' front, the ambition extended to reaching out beyond those who were involved in STD programme activities directly. By systematising and disseminating a successful methodology along with any transferable lessons learned, it was hoped that the programme might leverage its own impact and influence innovation processes more generally.

The scope, ambitions and aims of the STD programme—strategic and inter-sectoral—implied that the programme should be co-sponsored by a consortium of

4   This is in line with the principle of shared responsibility for sustainable development. It is also likely to lead toward more robust, better-designed and better-accepted solutions, which increases the chance of reaching a more sustainable society.
5   Analogous to the 'think globally, act locally' *leitmotif* of sustainable development, the sustainable technology development *leitmotif* might be 'think in jumps, act in steps'.
6   The role of any government programme should therefore be restricted to setting challenging, long-term eco-efficiency targets and to playing a facilitating role in the innovation process by helping to define innovation challenges, helping to set up appropriate innovation networks and providing technical support for these until a solution direction and an action plan are agreed by its members.

ministries rather than by any one alone and that funding should be sought, also, from private sources. In the event, five ministries provided block funding for the programme for an initial period of five years (see Chapter 1). Private financial backing was secured on a project-specific basis. The day-to-day management was entrusted to a programme director reporting to a supervisory steering committee made up of representatives of the participating ministries and a scientific advisory board. Business and industry input to programme steering was achieved via input to the steering committee and through project-specific industry forums. The STD programme was also supported by a consultative forum made up of representatives of societal organisations and stakeholder groups whose role was to encourage a wide societal awareness of the need for sustainable technologies, to bring stakeholders' concerns to the attention of technology designers and to build constituencies in support of promising technological directions.

The programme director was responsible for establishing a bureau, which was charged with carrying out the day-to-day activities of the programme, including:

▷   Developing the guiding methodology

▷   Establishing a set of case-study projects to test and refine the methodology and to initiate new technological trajectories

▷   Providing technical support and facilitation services to the new innovation networks

▷   Overseeing the separate innovation networks to identify common interests and synergies

▷   Liaising with political representatives about the structural conditions needed to promote solutions revealed through the case studies

▷   Studying conditions implicated in the success/failure and survival/collapse of networks

▷   Monitoring innovation experiences to draw lessons from the case studies about the method

▷   Communicating and disseminating information about the programme and its methodological and substantive findings, nationally and internationally

THE THEORETICAL BASIS OF THE PROGRAMME

The central idea of the STD programme is that the prospects for sustainable technology development could be improved by manipulating or modifying innovation processes. In seeking ways to influence innovation, the STD programme drew on several theoretical and methodological propositions deriving, especially, from social network theory and experience with backcasting methodology (Box 4.1). These are that:

▷ Social networks are key elements both in the stabilisation of present technologies and, potentially, in the creation of new ones.

▷ Shared problem definitions constituted on a 'first principles' basis (starting with a redefinition of the need to be fulfilled) are potentially key elements in the creation of new innovation networks and new technological trajectories.

Box 4.1 **The theoretical and conceptual basis of the STD programme**

TECHNOLOGICAL INNOVATION HAS BEEN ANALYSED IN TERMS OF SOCIAL NETWORKS that generate and support a technology (Vergragt 1988; Mulder 1992). Each network and the technology that it develops become mutually self-sustaining in terms of identities and rationales. Social network theory offers insights into the autonomous, self-perpetuating nature of the innovation process. Its proponents argue that, as a result of the influence of innovation networks over innovation processes and because of the survival instincts of network members, evolutionary development rather than technology renewal is the innovation norm. R&D resources are allocated accordingly. The considerable significance of social networks in technology dynamics provides one of the pillars of the STD programme approach. The central idea is that new directions for R&D might be found by creating new cross-sectoral networks around innovation challenges.

A key finding of social network analysis is the central role played by problem definition in the initial formation of a network of innovators. Critical decisions about start, direction, branching and termination of R&D activities are taken in groups or networks made up of individuals whose commonality is a shared problem definition. The networks and the problem definitions come, eventually, to define each other (Vergragt and van Noort 1996). Network dynamics relate strongly to the needs and requirements of the innovation process, which change as R&D shifts from one stage to the next (Mulder 1992). Ultimately, each network develops into a powerful constituency that perpetuates its technology. As well as creating new social networks, therefore, the challenge for the STD programme was to help network members redefine innovation challenges in new terms, especially by revisiting the nature of the 'need' that a technology is designed to serve.

Already since the 1980s, Constructive Technology Assessment has been used as a deliberate procedure in technology design to introduce an iterative feedback loop from users to designers (Smits and Leyten 1991). The purpose is to broaden the decision-making process and so increase the mutual adaptation of technological and societal developments. To achieve this, instruments such as 'consensus conferences' and 'future images' are sometimes used, where the objective is for a group consensus to emerge around a target for R&D efforts. Past applications of constructive technology assessment have mostly been oriented toward short-term social optimisation of existing technologies, for which the technique is well suited. However, the technique could also be applied in longer-term projects, with the aim of integrating long-term social considerations into technology designs from the outset.

▷ Creativity can be stimulated by proposing challenging eco-efficiency targets and drastic discontinuity: for example, by asking what would be possible in a world without fossil fuels.

▷ Backcasting is a possible tool for establishing shared visions of desirable future system states and for securing a 'systems' perspective on the transition process. It can also help in defining feasible short-term actions that can lead to trend-breaking change.

▷ Sustainable technologies will need to meet multiple criteria pertaining to future acceptability and viability. Constructive Technology Assessment can offer opportunity for participatory development of sustainable technologies and improve the chances of securing co-evolutionary innovation and paradigmatic change.

▷ The environmental impacts of potential solutions should be evaluated on a life-cycle basis against the benchmarks of today's technological solutions and the target of achieving factor-10 to factor-50 reductions from anticipated future claims on eco-capacity.

▷ Prototypes of sustainable technologies will need to be tested and refined in sheltered conditions in the context of specially selected and protected market niches.

The essence of the STD programme approach, then, lies in the mutually supportive use of need-driven approaches to problem redefinition, backcasting methodology and new innovation networks to explore the challenges to technological development posed by sustainability. Central to the approach is the conceptualisation of long-term innovation as a social process that involves interaction among members of an innovation network. The starting point for the creation of a new innovation network lies in redefining an innovation challenge from first principles in relation to a key area of future need. The expectation is that sustainable technologies will not be coincident with present sectors or solutions. The search process is therefore most likely to be creative and successful if the process is needs- or end-service-driven.[7] In each need field, a new network is built around a new definition of the challenge that the need presents for innovation and sustainability. The network is then encouraged to develop a range of promising new solution concepts and to bring at least one to the stage of an outline 'design' capable of being communicated to others. For this, network members should generate a shared expectation or vision about how the need might be met in the long-term future using sustainable technologies embedded in a compatible socio-economic context.

7   For the purpose of the programme, this suggests that the case studies should focus on needs that are relatively constant and predictable into the long-term future: for example, needs linked to basic requirements such as for nutrition or for energy services.

When developed, new technologies are often first introduced into strategically chosen niche markets where they can be refined in test situations away from the threat of market competition. The niche into which they are introduced is often established and protected by participants in the relevant innovation network (Rip 1989; Schot 1992). Within it, the technology and its users have opportunity for mutual adaptation and, through experience, designers have opportunity to improve the performance of their designs, adapt them to users' requirements and lower production costs. By helping to establish protected niches for new technologies, the STD programme could provide opportunities for long-term sustainable technologies.

By describing a pathway linking their vision of a target long-term sustainable future system state back to the present day, members of a network are able to define actions that should be taken in the near term to realise this vision. Because the pathways must be described through a logical sequence of coherent interim states, the traced pathways help to identify the main technological, economic and behavioural challenges that will be faced and by when these should be overcome. Moreover, since each pathway is drawn right back to the present time, the first actions will be ones that can be taken today. By implication, each pathway can be converted into a descriptive 'future history' that describes an evolutionary way forward toward trend-breaking technological and paradigmatic change. Back-casting, then, is the hallmark of the STD method (Figure 4.4). It is the main mechanism for building new networks, promoting creativity and incorporating a 'systems' perspective into the innovation process. It is used to relax unnecessary constraints on the creative process, such as to reduce the influence of existing

Figure 4.4 **Backcasting: a 'stepping stone' towards sustainable technology**

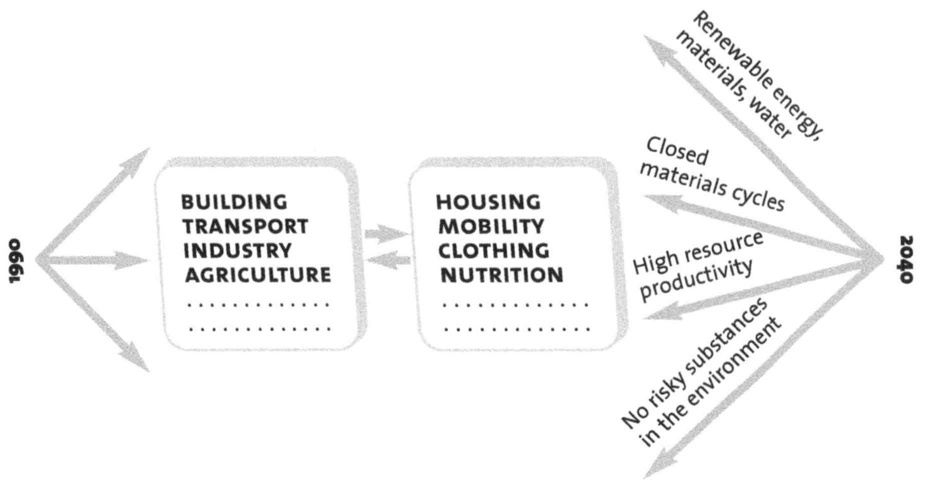

technological solutions, while reintroducing some necessary, but usually missing, sustainability constraints. Backcasting ensures that the challenge to technology developers is redefined from first principles and that the process is 'needs-driven'. This includes not only the function that the solution should deliver, but also the need for a systems solution and for a sustainable solution.

TOWARD AN OPERATIONAL APPROACH AND METHOD

As conceived by the STD programme, the innovation process therefore involves a set of tasks to be undertaken in sequence, each associated with one or more targeted outcomes. The STD programme seeks to support and influence the innovation process by providing a step-by-step working schedule, which clearly sets out the tasks and the anticipated outcomes, and by proposing tools, such as the backcasting methodology, which can be used to accomplish each task. The work schedule (Fig. 4.5) begins with an orientation analysis (first step) to see how

Figure 4.5 **The STD schedule**

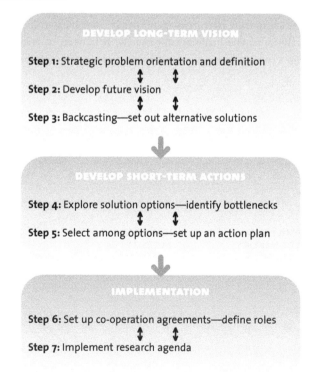

DEVELOP LONG-TERM VISION

Step 1: Strategic problem orientation and definition

Step 2: Develop future vision

Step 3: Backcasting—set out alternative solutions

DEVELOP SHORT-TERM ACTIONS

Step 4: Explore solution options—identify bottlenecks

Step 5: Select among options—set up an action plan

IMPLEMENTATION

Step 6: Set up co-operation agreements—define roles

Step 7: Implement research agenda

needs are presently met and to clarify the current and likely future sources of unsustainability if trends are unbroken. The sustainability challenge should then be redefined from first principles by revisiting the functions that an alternative solution should meet and other requirements, including the need to achieve a factor-10 to factor-50 reduction in the claim on eco-capacity. The opportunities and chances for new solutions should also be analysed, including possibilities opened by discoveries in basic science. Members of the innovation network should then make first charcoal sketches of their visions of how the identified needs might be met in new ways in the long-term future (second step). Thinking 50 years ahead gives creativity and inspiration a chance because the future visions need not be constrained by present arrangements. Backcasting (third step) reveals ways in which this vision of a sustainable future might be realised. Different possible solution pathways need to be further defined and evaluated (fourth step) in order to choose between alternatives and develop an action plan for tackling the technological and other bottlenecks that have been identified (fifth step).[8] A solid platform of support needs to be built around the action plan and responsibilities decided among parties to co-operation agreements and joint ventures involving 'innovation champions' who will carry the work forward (sixth step). The action plan for carrying the new innovation trajectory forward can then be implemented (seventh step).

The various steps in the schedule fall into three main phases of activity. The first phase (steps 1–3) is designed to develop a long-term vision based on a strategic review of how a need might be met in the future. The second phase (steps 4–5) is designed to clarify the near-term actions that are needed to realise this future. The third and final phase (steps 6–7) is concerned with realising the action plan. Although the schedule might give the appearance of a linear working approach, the working method is actually iterative.[9] This is depicted by the two-way arrows of influence in Figure 4.5. After network members have decided on a plan of action in regard to an R&D trajectory, the STD programme can also help to facilitate this: for example, by brokering co-operation agreements among innovators. For this, the programme can propose generic instruments, such as covenants, through which the parties are able to reduce uncertainty and risk by making binding commitments in respect to each player's roles, responsibilities and rights in the technology development process. Such covenants can be helpful in overcoming organisational obstacles to long-term innovation, which arise when operational

8   Often, it can be expected that the same technological, economic and behavioural bottlenecks will be faced in regard to many alternatives within the set of feasible solution trajectories. In this case, it may not be necessary to select any single solution trajectory as preferable to all others but, rather, to work to overcome bottlenecks that are common to several trajectories.

9   When dealing with the future, there are always questions and unknowns. This means that, whenever new knowledge is gained, there is need to take time to reflect and reconsider earlier steps in the work.

control is fragmented and different parties are interdependent in realising a new technological solution.

The core of the STD programme's activity involved testing and refining the approach using case studies in five key areas of future need and a set of illustration projects deriving from these, each exploring possible new solution trajectories. The key need-areas chosen as case studies were nutrition, mobility, urban living/working, water services and chemical/material services. Box 4.2 details the case studies and Table 4.2 lists the associated illustration projects.

ANTICIPATED OUTCOMES

The STD programme was intended to influence not only innovation processes, but also societal innovation contexts and corporate innovation cultures to make these

Table 4.2 STD case studies and illustrations

NUTRITION
▷ Novel protein foods
▷ High-technology closed-system horticulture
▷ Integral crop conversion
▷ Mixed land use systems

TRANSPORT/MOBILITY
▷ Urban underground freight transport
▷ Information technology for transport systems management
▷ Demand-responsive public transport
▷ Mobile hydrogen fuel cell

BUILDINGS AND URBAN SPACES
▷ Sustainable public housing
▷ Sustainable offices
▷ Urban restructuring

SERVICES PROVIDED BY WATER
▷ Differentiation in the municipal water chain

SERVICES PROVIDED BY MATERIALS/CHEMICALS
▷ C_1 chemistry
▷ New chemical engineering approaches in fine chemistry
▷ Cascade use of biomass
▷ Structural materials from natural-fibre composites

NUTRITION

The production of a single kilogram of pork takes 4–5 kg of feedstock. Energy, materials and land are wasted on a large scale. Also, the production of other foodstuffs is inefficient. The more mouths there are to feed and the wealthier the world's population, the more food will be required and the higher will be the quality demand in terms of variety, taste, freshness and nutritional value. We must find ways of supplying and distributing food that are more efficient and less damaging to the environment than today's production routes and, also, pre-empt the implications of restructuring for the agricultural sector and for those with economic interests in the human food chain. The Netherlands is a major exporter of food products, especially meat. It therefore has a deep interest in finding solutions to the challenges of sustainable nutrition.

MOBILITY/TRANSPORT

The length of the motorway network and the size of the vehicle fleet have become symbols of a country's economic and developmental status. Yet the condition and use made of the transport system can be ambiguous indicators. The demand for transport services is a function of spatial arrangements and behavioural choices as well as of the level of economic activity. The demand for transport and mobility continues to increase, putting strain on the physical transportation infrastructure, clogging up the distribution channels and generating unsustainable claims on eco-capacity. The transport of people and goods is very energy-intensive and puts multiple strains on the environment. The Netherlands already has the most intricate transportation infrastructure in the whole of Europe and, still, the system is under strain and is unsustainable. With the growing demand for mobility, alternatives have to be sought in demand management and in new supply arrangements that provide ways of moving freight and giving people convenient access to activities with much lower environmental stress. The Netherlands has considerable interest in new solutions to the challenge of sustainable mobility. As a major player in the world liquid fuel business, the Netherlands also has interests in future fuel systems and vehicle-propulsion technologies.

HOMES, OFFICES AND URBAN SPACES

People need buildings, not only to provide shelter but also to provide functional space for living, working and recreation. Today's buildings are eco-inefficient to build, maintain and run, and the building designs are inflexible. Half of all commercial energy is used in buildings. Meanwhile, renewable energy, falling as sunlight onto buildings, is wasted. Much of the potable water used in buildings is used for purposes other than drinking, such as for washing clothes, flushing toilets and carrying wastes away, while rainwater falling on buildings is channelled directly into the waste-water system. Today's architectural designs mean that living and working spaces are difficult to reorganise to meet the changing needs of families and businesses, which makes for a great deal of relocation and unnecessarily high levels of demolition and reconstruction. Today's urban arrangements are also eco-inefficient and often detract from quality of life for citizens rather than improve it. Urbanisation is taking place worldwide. Twenty-five years ago, 30% of the world's population lived in cities; now 70% live in cities. A significant sustainability challenge is to minimise the ecological pressure imposed by urbanisation while meeting people's living, working and recreational needs and enhancing the quality of urban life. The Netherlands, as one of the world's most highly urbanised societies and with a long-standing physical and urban planning tradition, is well suited to addressing this challenge.

Box 4.2 **The five STD programme needs-based case studies**

WATER SERVICES

More than enough clean water falls as rainfall every year, but uneven distribution and pollution threaten to create a serious water crisis in much of the world. A major inefficiency in the municipal water chain is that we use water to transport waste products. In the Netherlands, one-quarter of carefully prepared drinking water is used only to flush toilets and to carry wastes through the sewerage system. Sources of inorganic contamination within the infrastructure of the urban water system itself mean that waste-water treatment is neither efficient nor very effective and that organic material within the waste-water stream cannot be valorised as fertiliser because of persistent contamination. Producing potable water and treating waste-water create high energy demands. The challenge is to differentiate, separate physically and manage separately several different qualities of water to give a better fit between water service needs and the quality of the water used to meet these. A fundamental strategic review of municipal water management is needed to meet the challenge. This is a particularly appropriate case study for the Netherlands because of its specialised expertise in water management and its specific geographical situation.

MATERIAL/CHEMICAL SERVICES

We use materials to provide structural, functional and energy services in a wide range of application areas, making use of the physical and chemical properties of materials found directly in nature and of the possibilities of engineering desirable and useful combinations of properties into materials made by chemical synthesis. The demand for the services provided by chemicals and materials is increasing and, also, there is pressure to provide increasingly sophisticated products with specialised combinations of properties and 'built-in' intelligence. Such products of the chemicals industry tend to involve many process stages and the production processes are often materials-, energy- and waste-intensive. R&D in the chemicals industry is focused heavily on developing new products to meet high-value demands. However, from the perspective of sustainability, the long-term challenges lie in a combination of dematerialisation of the wider economy (using fewer materials to provide the same or better functionality) and in developing more sustainable source-to-service material chains. For the bulk chemicals industry, the challenge is to switch to renewable raw material sources and to reorganise downstream production processes so that these can make use of feedstocks produced from renewable or recycled raw materials. For the fine chemicals industry, the challenge is to increase the efficiency of conversions substantially, by using new reaction media, more efficient reactions, bio-technologies, batch processes and fewer process steps. For this, a fundamental reorganisation of production processes is needed, with chemical engineering solutions redesigned specifically for the fine chemicals sector rather than, as now, modelled on those of the bulk chemicals sector. The Netherlands has a substantial chemicals sector and considerable expertise in the areas of chemistry and chemical engineering. Finding solutions to the sustainability challenges posed to the chemicals sector is important for Dutch industry and society.

Box 4.2 (continued)

more favourable for new solutions. These wide-ranging aims and objectives were matched by a wide-ranging set of anticipated outcomes:

▷ A first set of illustrations of technological solutions that could contribute to meeting future needs on a sustainable basis, described along with their cultural and structural context

▷ A more detailed definition of the technological challenge of sustainable development, of the technological possibilities for meeting that challenge and of the conditions on technological solutions[10]

▷ A set of 'embedded' innovation lines—continued independently by privately financed and managed R&D consortia—in respect to at least half of the illustrations begun through the programme

▷ Spin-off innovation lines begun and continued independently within Dutch companies and institutions without any direct programme involvement, but influenced indirectly by the programme and its methodology

▷ A greater appreciation in society of the importance, magnitude and urgency of the challenge of sustainable development, of the shared responsibility for sustainable development and of the role of sustainable technologies in providing solutions

▷ A clearer articulation by members of society of the demand for sustainable technologies

▷ A change in corporate culture to perceiving sustainability as an important opportunity, not a threat

▷ A method, guiding manual and transferable lessons for sustainable technology development

▷ Training materials and trainers with experience gained on the illustration projects by 'learning-by-doing'

▷ A set of international collaborations on long-term innovation processes/projects for sustainable technology development

---

10   This is important to provide a perspective on the programming and prioritising of R&D for business, technical institutes and universities, and a basis for the programming of policy.

FINAL REMARKS

The remainder of this book is devoted to describing the programme's work, methods and findings, using the case studies as a framework for organising the review. Since our major purpose is to describe the STD methodology for influencing innovation processes we make no attempt to cover the entire programme, which is too extensive to report here. Rather, we have taken a selection from the set of available materials, using specific case studies and illustration projects to highlight particular methodological points or steps in the approach. Part 2 of this book begins, however, with a chapter that uses the STD case study on nutrition to show how the various elements of the approach were integrated into a single, coherent methodology.

Part 2

Chapter 5

The STD process exemplified:
the case of nutrition

This chapter draws on the work of the STD sub-programme on nutrition and, especially, on reports produced by Oscar de Kuijer, Jaco Quist, André de Haan and Hans Linsen (Quist *et al.* 1996).[1] The STD research on nutrition makes a particularly useful case study from a methodological perspective since, of all the need fields analysed, this work most closely and completely followed the idealised sequence of steps described in Part 1 of this book. Also, the research on nutrition followed several lines of innovation and work on one or another of these lines carried on throughout the whole five years of the STD activity. The work on nutrition, therefore, offers a useful illustration of the learning-by-doing approach that ran through the programme as a whole. Lessons learned through work on earlier projects were applied here and lessons learned here were carried over into other fields of analysis. Another important reason for looking closely at the work on nutrition is that it provides good examples of some of the key characteristics of the search for breakthrough technologies.

As to the last of these, we can count three key characteristics. First, many of the individual technologies that the work identified as being important for achieving

---

1   The nutrition sub-programme was co-ordinated by these members of the STD team. In addition, work for the sub-programme was carried out by TNO-Leiden, TNO-MEP, ATO-DLO, LEI-DLO, the Wageningen Agricultural University, the Centre for Environmental Studies at Leiden and the SWOKA Institute for Consumer Affairs. The firm of consultants, Arthur D. Little, was commissioned to undertake work in the orientation phase of the programme. Ivo Larsen of Arthur D. Little contributed to the writing of one of the main reports. The Ministries of Economic Affairs, Agriculture, Nature Conservation and Fisheries, Public Health, Physical Planning and the Environment co-financed the work. In addition, sponsorship from industrial participants was received from Unilever, Gist-brocades and Avebe.

breakthroughs in the field of nutrition are the same as those needed to achieve breakthroughs in other need fields: for example, photovoltaic cells, sensors and optical fibres, all of which feature here, also feature in projects related to sustainable chemicals, materials, energy and washing. Second, leaps in eco-efficiency in this need field, as in most others, will not come from one technology alone. The technical possibility of making major jumps depends on synergistic clusters of technologies brought together to achieve specific purposes. In regard to this, the possibility of developing 'high-tech' closed-cycle horticultural systems that harness solar energy as the key renewable resource and depend on a technology cluster embracing at least six breakthrough technologies is a typical case in point. Third, the need for a co-evolution of technological, structural and cultural innovations to realise the technical potential of such opportunities is also well illustrated by this sub-programme. The exchange of ideas and information between research teams and societal stakeholder groups and the mutual learning that this sub-programme fostered exemplify that successful innovation depends as much on successful social processes as on successful laboratory work.

ORGANISATION OF THE WORK

The work on nutrition was organised in a series of seven steps, broadly representative of the idealised organisation discussed in earlier chapters. These steps can be grouped into three phases: a first 'orientation' phase aimed at defining problems, opportunities and possible solutions, identifying stakeholders and pinpointing evaluation criteria; a second 'definition' phase focused on refining and selecting among ideas; and a third 'implementation' phase focused on the organisation of work to carry forward a preferred idea or set of ideas. A fourth phase of demonstration and development, mostly concerning tasks beyond STD's immediate mandate, is the logical follow-up to a successful R&D exercise in the implementation phase. In this chapter, we do not go into the details of the demonstration activities that followed the work on nutrition. However, it is worth noting that this was one of the most successful of STD's activities and independent innovation agents have since carried forward every one of the innovation lines started here. In effect, all the innovation lines explored in the course of the work on nutrition have now become embedded within Dutch society and are being progressed without further STD involvement or support.

Within the nutrition sub-programme, the first step in the orientation phase was to make an initial analysis of the Dutch food sector to establish the main features of food production, consumption and trade and to pinpoint the major unsustainabilities associated with it. This orientation analysis was also intended to reveal where the major inefficiencies in food production and consumption lie, where opportunities to improve eco-efficiency might be found and what the main technological and other bottlenecks to taking up these opportunities might be. It was also an opportunity to identify relevant stakeholders. The orientation analysis was broad-ranging because it needed to cover both the production and consumption sides of nutrition. Of necessity, this meant sector-wide analysis, with consideration for the full value-adding chains, including all processes and activities that take place in the Netherlands irrespective of where consumption of produced foods takes place. A simple need analysis was also performed. Current trends in agricultural production, diet and values were used to provide guidance about the likely future environmental situation under a 'do-nothing' scenario.

In the second step, these aspects were drawn together to sketch a vision of a sustainable future as the basis for backcasting. Workshops were held among stakeholders to facilitate this. The future vision fed into a backcasting exercise, which was used to identify possible future directions for developments in food and nutrition. In this third step, several possible trajectories were identified together with the opportunities that each gave for radically increasing eco-efficiency. This provided the basis for evaluating the desirability of these trajectories and clarifying what would be needed to implement them, including what technological and other bottlenecks would have to be overcome. During the first three steps, a stakeholder analysis was used to identify and structure the criteria that stakeholders consider are relevant to the design and evaluation of future development trajectories. In a fourth step, a workshop was held to give direction to research by identifying promising themes for reducing the environmental burden of the present and future nutrition system. More detailed ideas about possible ways of meeting future nutritional needs and the technological challenges of these were tabled during a creativity workshop. This focused on identifying opportunities and bottlenecks, which were then used to structure a small set of research themes.

Moving to the second, definition, phase, the major task was to work out each of these themes in detail, exploring more fully both the technological breakthroughs needed and the organisational and practical aspects of implementation. These included four self-contained themes: multifunctional land use, high-tech agriculture, integral biomass conversion and novel protein foods. In addition, a fifth, cross-cutting, theme was identified concerning the development of sensor technologies that are needed for each of these four themes as well as in other need fields. In each case, the aim was to develop a research agenda that would help fill gaps in knowledge, experience and technological capabilities and to arrive at a communicable design of a needed breakthrough innovation. The objective was to set in motion an iterative and inclusive process, involving actors (those capable of developing and implementing technologies) and stakeholders (those who will use or will be affected by new technologies and whose attitudes and behaviours will be important for successful diffusion), that would ultimately lead to an agreed action plan for implementation. Within this phase, the fifth and sixth steps—working out the ideas and building a platform of support among stakeholders—occurred more or less simultaneously.

The third phase builds on the outcome of the second. It involves finding one or more actors who are willing to take the lead in 'championing' the best ideas that have emerged and, for which, a platform of support and approval from a stakeholder network has already been built. For STD, the third phase involved working closely with and supporting such 'champions' and adopting a facilitating role: for example, by helping to co-ordinate the new networks of actors, by brokering joint financing arrangements or by locating appropriate technical and other forms of support for the needed R&D. In the seventh and final step the objective was to help the newly formed innovation network and its leaders to launch a concrete plan of action for R&D and innovation that could be carried through in the future without further STD support.

The rest of this chapter is structured around a detailed description of each of these seven steps, using the nutrition sub-programme as a case study. Since the orientation phase (steps 1 to 4) was common to all five of the projects carried out on nutrition, this is described at the level of the sub-programme as a whole. A brief description is given of all five of the research themes identified in step 4. This chapter then follows the progress of two of these—integral biomass conversion and high-tech horticulture in controlled closed-cycle conditions—to illustrate steps 5 to 7. Some of the other themes are treated separately in other chapters. Chapter 5, for example, describes the work on novel protein foods.

STEP 1:
ORIENTATION

The orientation analysis is aimed at providing a picture of the present situation and an indication of what current trends might imply, especially for the nature and extent of the environmental problems faced in meeting nutritional requirements in the future. A breakdown of current nutritional patterns in the Netherlands shows that ten product categories constitute more than 90% of the total diet (Fig. 5.1). The figure reveals the strong contribution of the meat and dairy groups to total energy, protein, fat and carbohydrate consumption in the Netherlands. Proteins and fats tend to be over-represented in the current Dutch diet relative to the recommendations of the major health organisations about the ideal dietary balance, whereas carbohydrates are under-represented (Fig. 5.2).

Using the set of critical environmental themes identified by Weterings and Opschoor (1992), the current eco-capacity requirement associated with the production and consumption of each of the major food product groups was analysed and

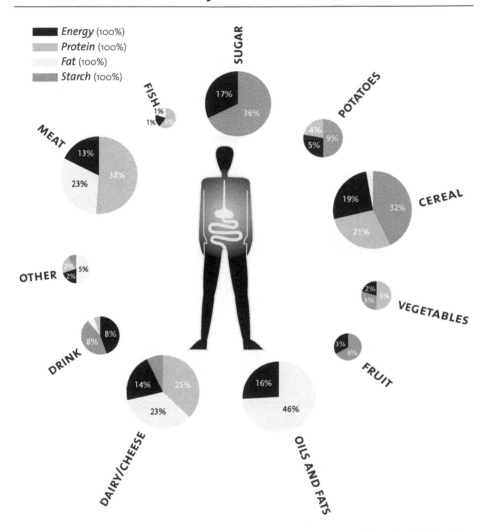

Figure 5.1 **Relative contribution of the major food product groups to daily nutrition in the Netherlands**

the products were ranked according to their absolute and relative unsustainabilities. Figure 5.3 shows that environmental concern is particularly high for meat, dairy, glasshouse vegetable and potato production. The major unsustainabilities from meat production arise because of the intensive rearing of livestock, which leads to problems in disposing of animal wastes. The use and run-off of macronutrients in open-field agriculture (nitrogen, phosphorus and potassium) is also an environmental problem, as is toxicity caused by the use of herbicides and pesticides. In addition, land use emerges as a cross-cutting concern because it is

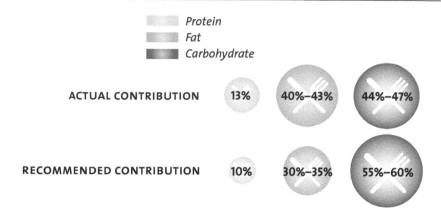

Figure 5.2 **Profile of the average Dutch dietary intake of proteins, fats
and carbohydrates in comparison with the recommended profile**

implicated in all current food-producing activities. Although this might seem
inevitable, this is related to current technologies of food production, which are
mostly soil- and land-based. Some alternative methods for producing conventional
and unconventional foods (for example, via biotechnological routes) could, in
principle, offer the possibility to substitute energy and water for land in food
production.

Box 5.1 gives results culled from trend analyses of consumer behaviour, showing
how ongoing changes in eating habits are likely to affect what is consumed in the
future (de Haan *et al.* 1997). There is a gradual shift away from carbohydrates,
especially potatoes. This is linked to a move away from the traditional main-course
meal, which, in the Netherlands, would typically feature potatoes as the starch.
Instead, people are eating more exotic meals—perhaps reflecting wider horizons
opened up by migration and foreign travel. Society is becoming multicultural;
Indian, Chinese and Thai foods are becoming more popular. At the same time,
there is a shift away from traditional eating routines. Both the shift to more exotic
meals and the emerging 'snacking' habit mean that consumers now tend to eat
more ready-prepared foods than before. These trends are expected to continue as
generation cohorts advance through the population and new habits become
increasingly dominant through the natural mechanism of population turnover.

The trend analysis suggests the need for some reinterpretation of the environmental issues. Problems associated with potato production, for example, will
probably diminish owing to reductions in potato use. Improvements in potato
production methods through research on pest-resistant strains of potato and better
targeting of pesticides used will reinforce this conclusion. As a consequence, potato
production is unlikely to continue to constitute a major environmental problem.
Importantly, the orientation analysis also looked at patterns of food trade, as this

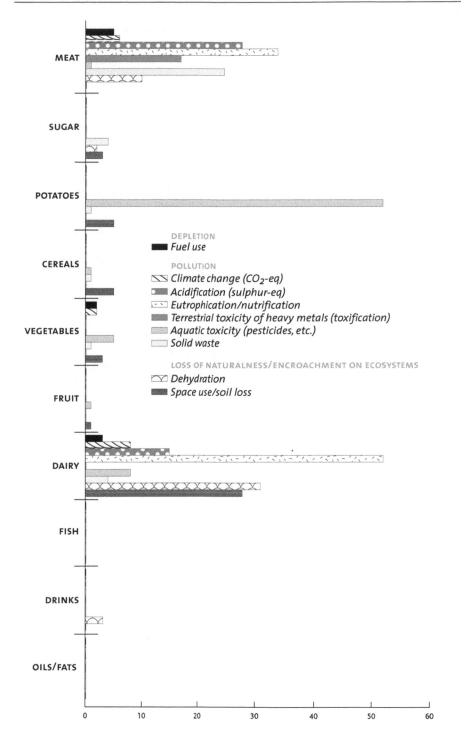

Figure 5.3 **Environmental load associated with each of the main food groups as a percentage of the total environmental load**

SURVEYS REVEAL THAT EATING HABITS ARE CHANGING. ALTHOUGH MEAT RETAINS a prestige association, the overall trend is toward a gradual decline in meat consumption. This is linked to an increasingly negative image among some consumers about meat production processes (arising from safety, environmental and animal rights concerns) and to concern for healthier diets. Another factor is the limited choice of meat variants. In contrast, there is a compensating trend toward increasing consumption of fish, which has a better public perception in terms of production process, health image and range of product choices. The overall consumption of dairy products is constant, although there are some changes in the structure of consumption: for example, cheese consumption is declining. Grain consumption demonstrates an increasing trend, whereas potato consumption is falling. In part, this reflects a shift away from the traditional meal of meat, potatoes and vegetables. Vegetable consumption remains constant overall, despite the trend away from the traditional meal, because there is a compensating shift toward vegetarian foods and healthier eating. Fruit consumption is also constant. The overall consumption of sugar in the diet is constant and is predicted to remain at today's level despite some changes in the structure of sugar consumption. There is an increase in the consumption of polysaccharides, oligosaccharides and invert sugar and a decline in unrefined sugars. The same holds for oils and fats.

As well as changes in the composition of the diet, eating habits and behaviours are changing in other ways too. Consumers are demonstrating a preference for continuous, day-long consumption of many small meals and snacks rather than for three large meals spaced at intervals throughout the day. Eating is becoming more closely tied in with other activities that are performed simultaneously, such as working on PCs, watching television and doing homework. Eating out is also part of this trend of combining eating with other activities, both work and leisure. Eating out has grown in importance and this trend is expected to continue in the future. Whether eating at home or in restaurants, the growing tendency is to eat meals that are already prepared by others. At home, the tendency is to eat more ready-made meals and snacks rather than to prepare meals from original ingredients. It is expected that future consumers will be inclined to 'graze' continuously on prepared foods and snacks. This will tend to increase the 'processed' share in the protein market, such as for ground beef rather than steaks or roasts. Nonetheless, consumers are still likely to retain a few traditional meals, since eating fulfils more than simply a nutritional function. The traditional weekly family meal is expected to survive long into the future, playing a mainstay role in family affairs as well as in family nutritional requirements. Meat is likely to retain a prestige association in this role.

Another important trend is the proliferation of factors that influence consumers' eating choices and the need this generates for information about foods and food production processes. Health, environmental and ethical aspects are playing an increasing role in consumers' eating choices. It is expected that future consumers will be more informed about the nutritional implications of their personal genetic make-up and lifestyles and that they will seek a better match between their individual nutritional needs and diet. Increasingly, consumers will want foods to impose lower environmental burdens and social costs. These aspects will join the traditional criteria of quality, taste and value for money in eating choices.

Box 5.1 **Value migration in nutrition**

also has a bearing on trends in environmental burden. Using trade data, the relative contribution of domestic and foreign nutritional requirements to the unsustainabilities revealed in Figure 5.1 can be calculated. It emerges that the environmental burden of potato production should decline because potato production within the Netherlands is mostly for domestic consumption, which is declining. In contrast, a large share of meat production is for the export market. Overall, around 70% of total Dutch food production is bound for export markets. The environmental stresses felt within the Netherlands are therefore less a consequence of meeting domestic needs than of supplying foreign markets. By implication, a major impetus for technological innovation in this sector is to find ways of reconciling environmental objectives with the objective of maintaining a strong export industry.

STEP 2:
SKETCHING A
FUTURE VISION

Stakeholders in the nutritional system— including those involved in today's food-producing system, possible alternative food-producing systems, consumer organisations, environmental groups, research institutions and government agents— were invited to participate in a scoping workshop. The goal was to paint a broad picture of a possible future food system and its configuration. As well as analysing the present and prospective future food and nutrition system and its environmental implications, participants looked for the major inefficiencies and the opportunities for reducing these at a strategic 'systems' level.

Several major sources of problems were identified. First, agricultural production on open land is, at present, highly specialised. Animal husbandry and crop production are mostly achieved on specialised farms devoted to one main product only. As a result, food production is poorly integrated into the natural ecosystem. Natural ecosystems are reduced and destabilised by monoculture and are unable to provide a continuing and sufficient stream of service functions for sustainable food production such as natural nutrient fixing, pest control, water supply and waste processing. Instead, farm production becomes dependent on artificial inputs, which are generally only poorly targeted. Specialisation also means that wastes produced on farms cannot easily be used productively as inputs to other processes. Waste-streams are generally so large and so spatially concentrated by monocultures that they outstrip local eco-capacities to deal with these. The concentration of pig rearing in the Netherlands is a special case in point. Dutch pork production is based on imported grain feedstock, such as soybean imported from the Far East. Much of the meat production is for export markets, but the animal wastes are left behind in the Netherlands in concentrations that would never arise naturally because the Netherlands could not itself produce the necessary grain feedstock. More generally, specialisation and concentration also imply that production occurs at distance from markets. This requires additional functions such as storage and transportation, which in turn pose extra demands on resources, especially energy.

This last source of environmental problems applies also to glasshouse production of salad crops and vegetables. At present, production is often far from points of consumption. Demand and supply are poorly integrated so that a production level is needed above actual consumption level in order to provide choice and ensure that all wants are met. A percentage of production is never bought or consumed. Further losses—around 10% of total production—occur during storage and transport. In addition to these inefficiencies, arising from logistical arrangements, present-day food production in glasshouses is far from efficient in terms of the conversion of inputs to useful outputs. In principle, solar energy is sufficient to meet the needs of plants for both heat and light. In today's practice, however, additional fossil energy is needed to compensate for the intermittent availability of solar energy. Similarly, in principle, there is a need to replace only the water and nutrients that leave the glasshouses within the produce. Yet, today, more than ten times these amounts are added because of losses to the environment.

Solar energy, water and carbon dioxide (together with genetic material) are the key resources for food production since these are the elements needed for plant photosynthesis. However, most sunlight is lost to plant production because it is received in urban areas or in places where there are topographic, soil or water constraints on plant production. Moreover, plants are able to capture light energy only when this is received at wavelengths between 400 and 700 nanometres. Less than 40% of received solar radiation is within this range. More generally, only 1%–2% of incoming solar energy is fixed through plant photosynthesis into sugars and carbohydrates. Plants vary in their efficiency in fixing solar energy, but even the most efficient, such as sugar cane, achieve a conversion efficiency of only a few per cent. Another major inefficiency is that only a small fraction of plant biomass is directly edible. The majority of primary production takes the form of biomass that is not eaten (plant stems, leaves, seed and fruit husks and peel, etc.), but which may nonetheless contain valuable oils, fats, nutrients and vitamins. In practice, less than 10% of plant biomass is represented by those parts of plants that are eaten as foods, such as edible fruits, seeds, seed-pods, roots and tubers. The rest is usually returned to the soil after composting as a soil conditioner. All of these imply that the basic renewable energy resource needed for food production—solar energy— is used very inefficiently. The energy content of food crops represents probably less than a tenth of 1% of the incoming solar energy received even within areas devoted wholly and specifically to agricultural production.

This inefficiency in primary production of plant biomass from solar energy is compounded if captured energy is then passed through an animal converter before being eaten by humans. From each trophic level of the food-chain hierarchy, only a small fraction of fixed energy, around 10%, is passed on to the next-higher level. In the case of meat and dairy production, for example, 5–10 kg of grain feedstock is needed to produce 1 kg of meat or dairy produce. In the case of cellulosic feedstock such as grass, which humans cannot directly digest, the use of an animal or other form of converter may be needed for humans to access the fixed energy. In the case of grain feedstock, however, a huge inefficiency is introduced by the preference of human consumers for meat protein. In the Netherlands, the emphasis on pork production using imported cereals from the Far East as a feedstock — such as soy from Thailand—is a case in point. For this reason, Dutch meat production is one of the most unsustainable of all economic activities since it relies heavily on the 'importation' of overseas eco-capacity in the form of the land and resources used to produce the imported feedstock.

On the one hand, these inefficiencies are the source of the environmental problems of the current system of nutrition. On the other hand, they also represent the opportunities for solving problems since they provide scope for improving efficiencies. Workshop participants saw several opportunities for this lying partly in new technological possibilities and partly in the already-discussed trends now affecting eating habits and nutritional 'rhythms'. In respect to the former, participants saw new opportunities opened up by the latest findings from basic research in such fields as plant physiology, genetic engineering, biotechnology, materials science, energy science and information technology. As to the latter, value migration toward healthy, environmentally friendly, exotic, ready-prepared meals and snacks was seen to offer opportunity for a new mixture of food ingredients and meals as well as new forms of food production to be welcomed and accepted.

The possibilities tabled by participants during the workshop can be arranged around several broad themes:

▷ Improved management and use of solar energy: for example, to ensure that more light energy is fixed as biomass and that, in controlled production environments, such as glasshouses, systems are in place to even out the variation in solar energy supply

▷ Closed-loop production: for example, in open-land farming systems to ensure that wastes from plants are used as fodder for animals and wastes from animals are used to supply nutrients for plants, and in controlled environments to ensure that losses of pesticides and nutrients to the environment are minimised

▷ Reduced use of artificial inputs: for example, by targeting nutrient, pesticide and herbicide applications in relation to needs at the level of the individual plant and by genetically modifying plants so that nutrient, pesticide and other requirements (including for solar energy) are lower

▷ Better integration of agricultural activities into natural ecosystems

▷ Better use of plant biomass so that less primary production is wasted: for example, to reduce production of inedible materials, losses during storage and transport, and losses during animal conversion

▷ Better integration of demand and supply: for example, to ensure that production takes place nearer consumption and is better matched to market demand

▷ Development of alternative or unconventional forms of foods that can supplement or substitute for inefficiently produced conventional meat proteins: for example, protein foods based on fermentation products or materials obtained from waste biomass

A broad vision of the future nutritional system began to be constructed around these opportunities based on a mixture of food production styles and food products. The elements of this vision included:

▷ High-technology production methods for production of conventional foods in contained, highly monitored and controlled, low-input and low-loss (closed-system) environments, with production occurring close to places of consumption so that transportation and storage waste are minimised and a better balance is achieved between demand and supply

▷ A multifunctional form of land use in open farming in which animal husbandry and crop production are integrated and combined with the use of land for recreation, energy production, water management, forestry, and wildlife preservation to give a diversified rural economy in which the economic functions are mutually supporting and fully integrated into natural ecosystems so that they reinforce rather than undermine natural ecosystem integrity, structure, function and resilience

▷ A full and integral use of produced biomass involving direct harvesting of edible plant parts with the remaining biomass used to produce other edible ingredients (such as oils, fats, vitamins and food additives) or as a feedstock for chemicals and energy production

▷ Development of complementary 'unconventional' foods, based especially on alternative (vegetable- or microorganism-based) forms of protein that could replace or substitute for meat protein (and could easily be incorporated into prepared meals and ready-made snacks as a replacement for minced meats and mechanically recovered meat pastes) and foods based on materials recovered from biomass wastes (such as oils and fats) that could substitute for dairy products

STEPS 3 AND 4:
BACKCASTING AND THE REVIEW
OF OPPORTUNITIES AND CHALLENGES

These ideas were constructed into a future vision in which the various aspects of changing consumer values and behaviour, the requirement to reduce the environmental burden of the nutrition system, and the possibilities opened up by new knowledge in several areas of basic science were blended into a qualitative picture of a plausible, viable and acceptable future. Five main themes were identified within this future vision, representing innovation tracks toward radically more sustainable ways of meeting nutritional needs. These were:

▷   Sustainable multifunctional land use

▷   High-technology closed-cycle horticulture

▷   Integral crop/biomass conversion

▷   Novel protein foods

▷   Sensor technology

The last of these was a technological theme that had already been identified as being of cross-cutting relevance for all the others. These themes were taken as the subjects for a set of parallel backcasting workshops involving actors and stakeholders.

Steps 3 and 4, backcasting and reviewing the innovation pathways that emerge as a result of the backcasting exercise, should more correctly be seen as interactive rather than sequential steps. The idea of backcasting is to trace back from a vision of a sustainable future in order to identify plausible pathways with which to connect that future with today's situation. In the backcasting process, technological trajectories and breakthroughs that can be started from the position of today's technological, cultural and structural contexts are identified along with a time schedule by which specific breakthroughs should be made if the vision is to be realised in time. Progress is made by means of an iterative process in which possible ways for connecting the present to a sustainable future situation are checked for technical, economic, environmental, social and cultural plausibility. An attempt is made to describe possible trajectories that imply consistent sets of technological, structural and cultural innovations and which will have a set of acceptable environmental, economic and social outcomes.

Backcasting can therefore serve many useful purposes. It can help to identify the range of stakeholder concerns over the acceptability of new technologies and over their impacts on, for example, established industries and jobs. This enables these concerns to be translated into criteria that can be used both to design and evaluate innovation trajectories. Backcasting can also be helpful in building platforms of support around promising innovation trajectories. It can help to reinforce the unsustainable implications of a do-nothing scenario and therefore makes a break away from current trends more likely. The potential costs and benefits of different courses of action as well as the conditions needed for successful innovation may also be more formally analysed using the backcasting methodology. Using back-casting it is possible to draw up a structured list of opportunities, bottlenecks and conditions for success across a number of potentially synergistic research fields and to see whether there might be complementarities between research and technology needs in several areas. Insights such as these can help change our perception of what long-term research is worthwhile. Especially, they can be useful in identifying key innovations which, if realised, would open up opportunities in several different fields of need. If the synergies are sufficiently large, some actors may consider working together to reduce the costs and spread the risks of R&D. In this way, actors may be convinced to commit resources to projects they would otherwise never have considered exploring. In turn, this can begin to relax some of the constraints on development pathways in other innovation fields and increase the probabilities generally of moving toward a more sustainable future.

Rather than describe the process in abstract, it is best reviewed in the specific context of the five research themes identified by the nutrition sub-programme, using these as examples.

MULTIFUNCTIONAL LAND USE SYSTEMS

In the case of open land, workshop participants considered that a multifunctional form of land use, in which several different possible ways of exploiting the same land area would be pursued simultaneously and synergistically, offers a more sustainable alternative to today's specialised and homogeneous exploitation regimes. Today, 60% of the Netherlands land area is devoted to agricultural production. Other major land uses are transport, leisure, waste disposal and water manage-ment. Today's highly specialised arrangements mean that economic functions are poorly—if at all—integrated into natural ecosystem functions. Moreover, few economic functions are mutually supporting. On the contrary, most are mutually antagonistic, imposing costs on and competing against each other. Specialisation adds to stress on the environment and to the economic costs of environmental protection. Under a regime of specialised land use, environmental protection is an add-on activity analogous to end-of-pipe approaches in industry. It depends, for example, on artificially providing services that would otherwise be provided free of charge by nature, such as nutrient recycling and water purification, and specifically setting aside protected tracts of land as nature reserves.

A multifunctional form of land use would aim at integrating several different forms of economic exploitation at the level of each land-holding enterprise. These could include crop and livestock production, timber and biomass production, energy production, water management, provision of leisure and recreational opportunities, habitat and landscape protection, wildlife and biodiversity protection, etc. As well as multifunctionality, the main elements of the proposed solution are a switch to decision-making at the level of a natural ecosystem (such as a drainage basin), greater integration of decision-making and activities among constituent farms and enterprises within the ecosystem, and more ecologically based criteria for designing systems of economic exploitation. These changes would enable modes of economic exploitation to be based on a set of ecological principles, such as maintaining and building ecosystem integrity, making best use of naturally provided ecological services, and closing the loops for water, nutrients and organic materials within the ecosystem. Closing the nutrient cycles has a double benefit since this reduces the need for additional inputs and reduces losses of contaminants to surface water. Lower levels of run-off and contamination make more surface water available for domestic use at lower production cost.

For these principles to apply, an exploitation regime for the ecological unit as a whole needs to be worked out together with stakeholders within the context of the specific local ecological situation and its past history of exploitation and environmental damage. Each land-holding enterprise would switch from a specialised activity to a multifunctional enterprise system. However, depending on its specific characteristics and location within the ecosystem, each enterprise would accent one function over others. Potential accents are on water, biomass and energy, on maintaining a cultural landscape, or on meat production. Wastes from one farm (such as manure from an enterprise producing pigs) would be used as inputs for farming activities on a neighbouring plot (such as an enterprise producing the fodder used to feed the pigs). All exploitation activities would be co-ordinated so as to maximise the economic value added and ecological resilience of the exploitation system as a whole.

To some extent, the implementation of this vision is an organisational issue. Nonetheless, for such a regime of exploitation to become possible, a range of technological challenges must be faced and overcome. In rebuilding natural ecosystems, or in duplicating their functions, a water table higher than that currently found on agricultural land needs to be restored and maintained and new agricultural regimes developed that are compatible with higher water tables. Reducing losses of nutrients and pesticides to ground and surface water is also essential if water production is to be an integral part of multifunctional land use. Also, new technologies are needed to enable the nutrients contained within organic wastes to be more easily assimilated by plants. Nitrates and phosphates produced by composting are not pure enough to be easily assimilated, so a high proportion of these nutrients are lost through run-off when compost is applied. Improved information about the nutrient status of soils and about the presence of disease vectors and pests is needed to reduce and better target agricultural inputs. These suggest that research is needed on several technological bottlenecks connected with closing nutrient cycles and with achieving multifunctionality.

As to the former, organic wastes from households should be used to make fertiliser so those nutrients leaving farms within foodstuffs can be returned. But, for this, the heavy metal content of organic waste needs to be reduced. Ways of preventing heavy metal contamination of wastes or of more effectively removing contaminants are needed, such as by separating the water streams for household sewage from those for storm-water. The problem of nutrient run-off also needs to be tackled if nutrient cycles are to be more tightly closed. Improving the assimilation rates and targeting of nutrients are keys to closing cycles and reducing losses. As an alternative to composting, thermal treatment of organic wastes may offer ways of increasing nutrient assimilation rates and needs to be explored. In addition, research is needed on sensors able to provide precise, real-time information about soil nutrient status, plant nutrient needs, and the presence and threat to plants from pest and disease vectors. Another important area of research (analogous to that needed on the return loop for nutrients between fields and households) is in respect to systems for feeding livestock using biomass waste and for returning manure back to feedstock growing areas.

As to multifunctionality, research is needed on the consequences for agriculture of having a higher water table. Research is also needed on rotation systems for enabling a wider variety of useful plant species to be grown. As well as conventional crops, rotations should include species (perhaps genetically modified) that are primarily intended for energy and materials production, such as poplars and miscanthus. For the production and storage of energy to be part of multifunctional land use, work is needed on technologies for capturing and transforming solar, wind and water energy (all as part of multifunctional land use), and on the potential use of groundwater in energy storage. Energy production can be integrated in the loops for nutrient recycling. In the case of pig farming, for example, digestion or fermentation can be used to produce energy from pig manure and organic waste. Waste heat from ventilation of pig units can also be recovered and re-used. But technologies are needed to improve the efficiencies of these energy recovery and transformation processes.

The potential benefits of such a multifunctional regime of land use were subsequently evaluated within this sub-programme in an actual case study developed in the agricultural region around Winterswijk. It was found that gains of a factor of 20 are achievable, in principle, across several important eco-capacity criteria (Fig. 5.4).

CLOSED, CONTROLLED PRODUCTION ENVIRONMENTS

In the case of production in closed, controlled environments, workshop participants looked for ways to retain the benefits of existing glasshouse cultivation of fruits, flowers and salad vegetables while minimising disbenefits. On the production side, a major concern is the fossil energy use of glasshouse horticulture. Today,

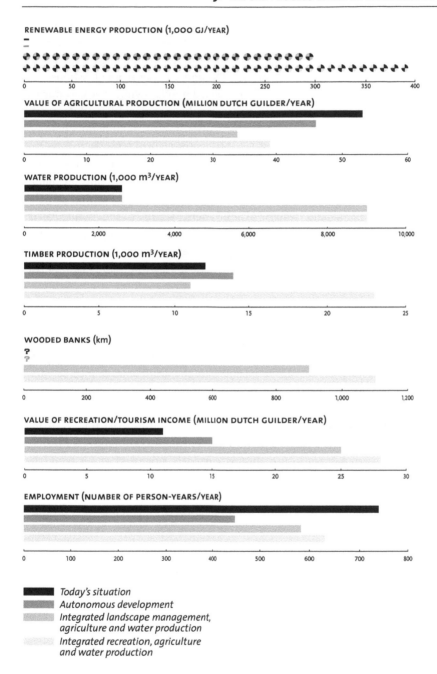

Figure 5.4 **Comparison of different single and multifunctional land use scenarios for Winterswijk**

natural gas is used to heat the glasshouses and maintain an artificially rich CO_2 atmosphere. Total natural gas consumption for glasshouse horticulture accounts for 10% of all current Dutch natural gas use. Other major concerns are wastage of solar radiation, losses of water and nutrients through escapes from glasshouses to the outside environment, and wastage of produced goods owing to losses during storage and transportation and poor co-ordination between supply and demand.

Since the total supply of solar radiation is, in principle, more than adequate for the growing needs of plants, the key to solving these problems is to use the available solar energy more efficiently. There are two technological aspects to this problem. The first is that the natural supply of solar energy is only available intermittently. The technological challenge is therefore one of evening out the diurnal and seasonal peaks and troughs of solar energy supply so that plant growth can occur evenly and continuously irrespective of the time of day or season of the year. A second aspect concerns the low utilisation rate of solar energy by plants. More than half of the received radiation is in wavelengths that plants cannot use for photosynthesis or falls in areas not used for plant production.

These problems call for technological solutions for making better use of solar energy. Plant physiology can be developed to improve rates of photosynthetic energy fixation. Semi-transparent photovoltaic cells that can capture solar radiation received at wavelengths not suitable for plant photosynthesis, but will allow the passage of light energy at wavelengths between 400 and 700 nanometres, offer a technological means of using solar radiation more efficiently. There are also possibilities for storing surplus heat energy during the day in groundwater and recovering it at night using heat pumps. Other opportunities exist for capturing and moving solar radiation using optical fibres. Through combinations of these energy technologies, solar radiation that is currently not used at all could be diverted for plant production.

But these energy technologies also offer scope for solving the other problems outlined above. Especially, better use of solar radiation would enable food production to take place closer to the consumer. It is even possible to think of high-technology-controlled production units being located within cities, making use of solar radiation that is received on rooftops and other solid surfaces and either transformed to electricity or moved directly to horticultural units as transported sunlight. The production units could then be locationally free from constraints imposed by the need for direct sunlight: for example, they could be underground or housed in multi-storey buildings.

The opportunities are even greater if a closed-loop form of horticulture can be achieved in which water and nutrients can be cycled. This depends on technologies essentially similar to those described for open-land production systems, especially on sensors and technologies for ensuring that the nutrient content of organic wastes from households is recovered and re-used and that the contained nutrients are made available in a readily assimilable form. The capacities to close the water and nutrient cycles and to move sunlight would, in principle, relax the key constraints imposed by water availability, soil quality and topography that currently prohibit food production in many of the world's difficult environments, such as in arid, saline and mountainous regions.

Technological breakthroughs are needed in respect to plant physiology, conversion of light to electricity, transport of sunlight, storage of energy, conversion of energy, and process steering and management.

FULL USE OF BIOMASS THROUGH INTEGRAL CONVERSION

Workshop participants focused on the opportunities offered by the larger fraction of biomass that is currently unused for nutritional purposes but which, nonetheless, contains nutritious and useful materials, such as proteins, oils, fats, vitamins, flavourings and colourings. In view of the fact that conventional livestock production (meat and dairy) is an inefficient way of supplying nutritional needs and imposes a high ecological burden, there is a potential double advantage if substitutes for conventional meat and dairy products can be made from ingredients derived from waste biomass. This opportunity is also consistent with trends in consumer tastes and preferences for healthier foods and with the migration of consumer value toward ready-prepared foods. These shifts in market preferences offer openings for alternative food products and ingredients produced from unconventional sources.

The concept of integral conversion of plants and biomass is that of complete extraction of all the useful ingredients contained within the molecular structures of plants formed during photosynthesis. The technological challenge is partly to increase the fraction and quality of useful molecular structures contained within the plant and partly to improve the separability of high-value components of the plant biomass from other material residues that might go on to be valorised in non-nutritional application fields, such as industrial materials or power generation. This depends on rethinking plant-breeding programmes, which currently focus mostly on producing high quantities and qualities only of the primary product or commodity. Several candidate crops and plant species contain high-value components or have potential to be developed in this direction. Flax and hemp are good examples. If the environmental problems associated with potato production can be solved—for example, by using genetically modified pest-resistant potato strains or bio-pesticides to reduce toxic run-off during potato production—this is also a potentially interesting and versatile candidate as a crop for whole and integral conversion.

A second technological challenge is to develop methods for extracting these high-value components in the field. This is necessary because of the weight loss involved in the production process. Conversion *in situ* is the only cost-effective way of obtaining secondary products from large volumes of solid biomass waste and, especially, of reducing the economic and environmental cost burden of transport by removing only high-value commodities from the field. This depends on developing small, flexible and precise facilities for biomass conversion so that processing can be carried out in a decentralised manner. Then, only the final products need to be transported away. The need is for technologies that can achieve economies of scope rather than economies of scale, as traditionally has been emphasised in technological research and innovation. It is possible that biotechnology offers such a route and that the search is for a small but effective bio-refinery whose flexibility is based on the use of different and highly specialised and targeted conversion organisms. These may need to be genetically engineered.

This suggests an emphasis on a co-ordinated programme of research to develop plant varieties and related conversion technologies, including the physical equipment (such as bio-refineries), supporting materials (such as bio-catalysts) and logistics systems (such as for transporting final products to processing plants) aimed at valorising biomass waste *in situ* through complete and integral crop conversions.

NOVEL PROTEIN FOODS

The same concerns that motivate fuller use of the nutrient potential of waste biomass also motivate interest in unconventional foods that could substitute for today's meat and dairy products. Alternatives to meat protein—for example, based on vegetable proteins and fermentation technologies—have already been in use for thousands of years. Several of the world's most traditional foods belong in this category, such as tofu, tempreh and gluten. Today, however, these alternative foods are only poorly represented within the overall diet of most people, and they are widely perceived as inferior in taste and texture to foods based on animal proteins.

Nonetheless, opportunities exist to develop a wide range of novel protein foods based on plant and non-plant proteins, such as genetically modified peas, beans, lupins, lucerne, bacteria (such as *Spirulina* and *Fusarium*) and combinations of these that have so far not been exploited. More than 50 promising sources of materials are already known. Many of these enable protein to be produced with factor-20 to factor-30 improvements in resource productivity and at substantially lower economic cost. Many incur less than one-half and some only one-fifth of the production costs of conventional meat proteins with equivalent food value. Moreover, if research effort was expended on developing food ingredients and products based on novel proteins so that these are as tasty and acceptable to consumers as meats, the opportunity exists to offer consumers a lower-priced and potentially superior alternative to meat, which is also much more sustainable in terms of the environmental burden associated with its production route. The value migration evident in changing consumer diets and habits, especially the shift toward hamburgers, sausages and products based on mechanically recovered meat and meat pastes, offers scope to introduce novel protein foods gradually in prepared foods and meals.

The major bottlenecks are social, cultural and economic rather than techno-logical. Consumer acceptance of novel protein foods and the socioeconomic impact of any major switch from animal to novel proteins (for example, on jobs and income generation in the traditional meat sector) are substantial concerns. This means that research on several parallel lines needs to be initiated and that communications and stakeholder involvement are especially important in build-ing an acceptable palette of technological options and a platform of support for progressing these toward implementation.

SENSOR TECHNOLOGIES

Across all of the research themes identified above, a cross-cutting technological need is for better information about growing conditions at the level of the individual plant (Kampers *et al.* 1995). Real-time information on the status of both the plant and its surrounding environment is needed to minimise waste and reduce all forms of intervention. This includes information on the presence of pest and disease vectors, light, temperature, the CO_2 and nutrient status of the immediate environment, and the precise condition and needs of the plant. Such information needs to be available continuously and to be coupled automatically with mechan-isms to process information and to deliver appropriate responses. Such sensory possibilities are now conceivable through developments in several fields (Box 5.2). A list of information needs and possible sensor technologies includes:

▷ Information on the chemical composition of complex and variable media, such as soils, manure, water and biomass, and the presence or absence of specific compounds. Likely technological solutions are electrical sen-sors, biosensors, optical-chemical sensors and microanalysis systems.

▷ Information at the level of an individual plant or animal on its condition and requirements. Candidate technologies are high energetic and nuclear techniques, spectroscopy, dielectric, radar and ultrasound techniques.

▷ Information about the surrounding environment, such as on the local microclimate, CO_2 concentration and light conditions. Candidate technol-ogies are physical sensors, such as light sensors, and optical and elec-tronic gas sensors.

▷ Information on the precise position, form and structure of plants in relation to their surrounding environment. Remote sensing and imaging (in two or three dimensions) are promising technological routes.

THE NEED FOR NEW SENSOR AND MEASUREMENT TECHNOLOGY IS ESSENTIAL FOR progress on each of the proposed trajectories for improving the environmental performance of agricultural production. For multifunctional (sustainable) land use and for full and integral crop and biomass conversion, both of which are open-field production systems, measurement technology is needed to control and optimise waste-streams as sources of nutrients, to monitor and control water supply, and for early detection of plant needs and health problems as means for minimising and better targeting interventions. For high-technology agricultural production in closed production units, measurement technology is needed additionally for automation and quality assurance.

CONTROL OF NUTRIENT AND WATER SUPPLY

To use organic wastes as a source of nutrients, these must have a texture and nutrient content corresponding to plant needs. Optimisation depends on being able to monitor and control the composition and mixing of different kinds of waste. Current procedures depend on off-line analysis; i.e. on sending samples to laboratories for analysis. The time-lags reduce the value of the results, since the situation being monitored is dynamic. A real-time (on-line) measurement and control technology is needed. The main technical complication is the complex composition of the waste matrix. Elements within it can be present in a variety of chemical compounds. Phosphorus, for example, can be present in its inorganic form or in various organic molecules. Today's analytical approach is to bring the elements of interest into a defined chemical state prior to profiling. A comparable on-line approach may be possible in the future (Gerben *et al.* 1994). An alternative solution is to use measurement technology based on nuclear or inner electron shell properties, which are not affected by the chemical state of the element. This technology (NMR and x-ray fluorescent spectroscopy) is available in the laboratory but would need to be significantly improved for large-scale, low-cost applications in fields.

To make efficient use of plant nutrients and prevent contamination of watercourses by run-off, nutrients need to be administered in quantities and qualities corresponding to plant needs and according to the water content of the soil at the time of application. High-frequency electromagnetic techniques could be used to determine, remotely, the amount of water and the concentration of ions in the soil. Chemical sensors, such as ion selective field effect transistors (ISFETs) may be used to measure the concentrations of specific ions (Gieling and van den Vlekkert 1996). Sensors may also be developed that determine nutrient needs from the plants themselves. Analogous to chlorophyll fluorescence, other probing techniques based on absorption or scattering of light in specific wavebands could possibly provide this information.

The amount of water available to plants is also important for nutrient control. Excess water will pass through the soil to groundwater, transporting nutrients with it and taking them beyond the reach of plant roots. Too little water is also a problem, since this inhibits plant development. Continuous monitoring and control of the soil water content is, therefore, critical. Control can be effected using artificial drainage systems, which prevent losses of excess water to groundwater and 'store' both surplus water and its nutrient content for re-use during dry-weather periods. Sensors are needed to measure both the water tension, which indicates when to take action, and nutrient content, which indicates what action to take. These measurement systems, such as dielectric water content sensors, must function reliably and reproducibly over a full growing season (Hilhorst and Dirksen 1994).

OPTIMISATION OF PRIMARY PRODUCTION

The primary aim of the project on integrated crop conversion was to find ways of converting agricultural inputs into food products in the most efficient way and with as little waste as possible. The two systems at the centre of the study are the cultivation boom, with ➡

Box 5.2 **Sensors for agricultural production in the 21st century** which primary agricultural production can be optimised, and the bio-refinery, in which microorganisms convert the primary production into a wide range of raw materials and products. Measurement technologies are needed to operate both of these systems, but are needed especially to operate the cultivation boom effectively and efficiently.

A cultivation boom may be devised to rotate over a growing 'crop circle' or to move linearly, back and forth across a rectangular production plot. Either way, because the boom moves over a well-defined area using tracks (and, in the case of the crop circle, has a fixed central point of rotation at the pivot of the boom), its position can be known and reproduced precisely if combined with an accurate position measurement and control system. Optimised interventions at the level of the individual plant are, therefore, possible, subject to relevant information on plant status being gathered by sensors mounted on the boom.

Pest, disease and weed monitoring and control are especially important to reduce interventions. Early warning of disease and targeted intervention can prevent disease and pest outbreaks from spreading across the whole crop. Interventions can include chemical controls (application of pesticides, herbicides, etc.), biological controls (application of natural biocides or biological agents) or mechanical controls (removal of the affected plant or offending weeds).

Sensors for early detection of disease or infestation may rely, among others, on colour changes, disfigurement, changes in fluorescence, changes in plant metabolism, and the presence of specific chemical substances (such as antibodies).

▷ Detection based on leaf colour changes can be made by comparing spectral information contained in reflected light monitored on consecutive passes.

▷ Changes in shape may be detected by systems based on computer vision, using colour and image information obtained in different spectral bands.

▷ Fluorescence behaviour can be measured at night to diagnose disease. Such diagnosis relies on the fact that certain disease-sensitive biological molecules within plants, having being exposed to light of a specific spectral composition and from a bright source, re-emit light after the source is switched off. The intensity and duration of the re-emitted light can indicate plant health.*

▷ Photo-acoustic techniques also offer the promise of being able to detect infestations, by measurement of chemical substances released in low concentrations into the air.

▷ The presence of insect pests may be detected by monitoring insect density in the air above the plants and, in some cases, also from specific noises.

▷ Diseases not detectable through these non-destructive tests may be monitored by harvesting plant tissue and soil samples and analysing these *in situ* on the boom.

▷ Weeds can be detected because of their different appearance and pattern of growth to crop plants and should be removed mechanically by boom-mounted tools.

The closed (controlled-environment) concept is aimed at drastically reducing non-renewable energy and agrochemical inputs of current glasshouse production systems. The concept is to obtain the energy for optimal growth conditions from the sun, recycle organic waste from the region as nutrients, and avoid chemical pest control. Closed-system production is difficult to invade for pest and disease organisms. Nonetheless, because air and water are recirculated (which increases the risk of spreading disease if this does enter the system), continuous measurement and control of infestations (using technologies of the same type as for open-field production) are needed. Sensing technologies are needed, also, for process automation and for product quality assurance. A specific need is to determine the precise position and

---

*   In the case of chlorophyll fluorescence, this behaviour is already measurable at close range (Snel *et al.* 1991). Work is needed to develop usable field sensors and to develop the concept to take in other forms of molecular fluorescence.

Box 5.2 (continued)

➡   status of leaves** and fruits to be removed or harvested. Measurement systems must be capable of determining the quality and ripeness of products with minimal damage before harvesting. Colour, shape and sugar content are ripeness correlates. Sugar content can be determined with near infrared (NIR) spectroscopy. Sensors for other quality parameters (weight, firmness, etc.) could be used to sort products automatically, accurately and objectively after harvest.

PROBLEMS AND POSSIBLE SOLUTIONS

Difficulties for technology development are introduced by biological variability, product fragility, the need for non-destructive testing, the large number of plants to be monitored, and the harshness of the environment in which sensitive sensor technologies must perform.

▷   Biological variability means that the relevant parameters to measure (for example, sugar content) and the relationships between these and the variable of interest (for example, ripeness) vary among different agricultural products. Measurement and diagnostic systems need to be developed that are product- (and maybe even cultivar-) specific.

▷   The relevant parameters are often attributes of living cells or of metabolic processes taking place within living organisms. Products are often fragile and easy to bruise. Sensors and measurement systems—whether remote or not—should not be harmful to plants or products. Contactless principles based on the interaction of electromagnetic radiation (light, microwaves, etc.) with the objects to be measured can reduce contamination and damage risks to growing plants and harvestable plant parts.

▷   It can be expected that, in controlled-environment systems, sensors will not be applied to all plants in respect to all parameters. The best results will be obtained if as many parameters as possible are monitored on a sample of plants and their information combined rather than by measuring all plants on only a few parameters. Systems based on smart sensors,[†] combined indicators and sample data are needed.

▷   The environment in which sensory devices must perform is neither high-technology-friendly nor laboratory-like. Many substances being measured contain aggressive compounds. Contactless principles are best, but not always possible. Microsystems based on micro-machining technology[‡] can provide solutions to measuring principles that require wet chemistry or recalibration of sensing elements. They can also enable sophisticated sensors and systems to be developed that can handle complex problems, but are small and can be produced at low cost.

**   To optimise the photosynthetic process, a well-defined leaf index must be maintained. Where light is insufficient, leaf growth is unnecessary and leaves should be removed.

†   Smart sensors combine sensing elements with micro-electronic circuitry for signal processing, calibration and data fusion (the combination of data from several sensors to give reproducible estimates for key parameters).

‡   These are electromechanical systems that combine small mechanical components for task performance, such as pumps and motors, with micro-electronic components for signal processing and system management.

Box 5.2 (continued)

STEPS 5–7:
FROM IDEAS TO ACTION

At this point in the process, the initial creative phase gives way to a more structured and formalised process of exploring, narrowing down and refining ideas. We enter the definitional phase of activities during which ideas are more fully worked out and a platform of support is built up around an agreed innovation goal, strategy and action plan. A design—sufficiently well described so that it can be communicated clearly to others—is carried forward into the final implementation phase, in which an action plan for converting the design from an idea to reality is carried forward by a stakeholder network and its leaders. For the purposes of exploring what is involved and exemplifying the process, we follow through these final two phases in respect to just two of the research themes that were identified by the nutrition sub-programme. The themes we use to exemplify the process are integral biomass conversion and controlled-environment horticultural production.

INTEGRAL BIOMASS CONVERSION

In following up this research theme, important links between the nutrition sub-programme and that on chemistry were forged. This is because there are many possible applications for materials contained within biomass if these can be extracted economically and in commercial quantities. Within the chemistry programme, a wide range of materials, technological routes and applications was investigated: for example, starch from potato waste as an absorbent material for babies' diapers, fibres from flax waste as a matrix for structural materials to replace steel, plastics and glass fibre in manufactured products, and chemicals that can be used as bio-pesticides. In the definition phase of the nutrition sub-programme, the emphasis was on evaluating one or two food products that might be made of ingredients obtained from biomass waste. This included cheese made of ingredients derived from lupins and pasta made mostly from starch obtained from potato waste.

As well as implying a methodological overlap between research in the two sub-programmes in respect to integral biomass conversion, the shared interest of the chemistry and nutrition sub-programmes in this technological route implies strong synergies that will help in the future development of the concept. In principle, biomass waste can be exploited for a wide range of contained molecules, especially complex molecules with sophisticated structures that are difficult and expensive to synthesise artificially through chemical routes. The greater the useful content and the more complete the use of the waste, the more cost-effective will be the exploitation, whatever uses are finally made of the extracted materials. Hemp, for example, can be used to provide a range of fibres and oils. The fibres can be used for textiles and paper production as well as for half-synthetic fibres (such as for synthetic silk production). The oils can be used to produce coatings, methanol and plastics. Importantly, it can be used for thermoplastics, since the oil is not saturated. Finally, the remaining waste can be used for clean energy production. In effect, the extraction of some compounds for use in foodstuffs and of others for use as materials is mutually supporting. Whatever use is made of the compounds, the biomass is being valorised and, because ingredients obtained from it can substitute for others, the overall burden on the environment is reduced.

In addition, because the approach suggests a shift in emphasis from a concentration on only one main product or commodity to several co-produced commodities, this route stimulates a new creativity in the search for commercialisable crops. Many plants that have never before been commercialised, and therefore have never undergone optimisation through selective breeding programmes and regimes of applied agronomy, are brought under renewed consideration. As a result, as well as currently farmed species and crops, there is also a growing interest in re-examining the large quantities of non-farmed biomass—such as seaweeds, trees and weeds—to see what materials might be contained within the biomass, what uses these might have and how they might be extracted. Again, a link with the chemistry sub-programme is the potential interest in plants that can grow on brackish soils and which may have little direct food interest—since they do not yield a directly edible crop—but which contain useful molecules, such as edible vitamins, antioxidants and oils or molecules with valuable industrial or pharmaceutical applications.

The need to shift toward using renewable materials together with the fact that so great a portion of produced biomass is currently wasted means that interest in 'whole-crop utilisation' is growing generally. Interest in biomass waste also reflects a growing realisation that limits are being reached to the scope for further improving the productivity of farmed plants and animals in respect to what has conventionally been considered the main commodity: for example, grains from the wheat plant, latex yields from rubber trees, coconuts from coco-palms. This is because many years of research effort have already been put into increasing yields of the main commodities and conventional routes for doing this have already been well explored and optimised. Such research includes selective breeding of plants and animals and the optimisation of the systems for producing these, including the development of cultivar-specific pesticides, herbicides, fertilisers and regimes for applying these. In contrast, very little research effort has gone into looking at the molecular content of the biomass wastes that are co-produced alongside the main commodity, even though these typically account for up to 90% of total biomass production. Equally, R&D has neglected plants that do not yield an easily or obviously usable commodity. Yet many 'weed' species do contain valuable molecules.

Against this backdrop, the project on integral crop conversion within the nutrition programme was concerned, in the definition phase, with describing and evaluating the potential of this technological trajectory more fully and in respect to concrete examples. At the same time, the purpose was to build a platform of support among stakeholders around an R&D programme that would be capable of taking the concept forward through to implementation without further support from STD. The need for technological innovations had already been vaguely outlined in the earlier phase. In particular, the need for *in situ* extraction of useful compounds from waste biomass and, therefore, for small and flexible extraction units had already been identified along with a potential solution, the bio-refinery, in which biological organisms, specifically developed for the task, would provide conversion and extraction services. Flexibility would be possible by changing the nature of the organism within the bio-refinery. In principle, this would provide a technological route for combining flexibility with small-scale physical equipment, avoiding the need for large-scale dedicated plant and transport of biomass waste. During the definition phase, attention was focused on how the logistics of such an operation could work in the field. Since the process would need to be very localised and the refineries would need to be small, the whole process depends heavily on automation, the existence of an intelligent control system and a very precise system of agriculture.

A concept was developed in association with the work on sensor technology, the cross-cutting parallel research theme. It is based on extending the use that is already made in intensive agricultural systems of rotating booms that deliver irrigation water and nutrients to growing plants. The concept is that these booms should be used also to mount sensors which, because they can be moved up and down the boom from one end to the other and will move over the plants during each sweep, can be used to obtain information at the level of each individual plant growing within the crop circle served by the boom. In principle, the boom can also be used to deliver interventions based on information received by the sensors and transmitted to a control centre, which could be local or remote. For this, the boom has to be equipped with appropriate devices (mobile along the boom like the sensors) capable of delivering the intervention. As well as delivering nutrients and pesticides, this could involve physical acts such as the removal of a pest, a diseased leaf, a leaf blocking sunlight from ripening fruits, or of materials ready to harvest. The booms would deliver harvested biomass to the bio-refinery, where it would be given a biological treatment. A single bio-refinery could serve a set of booms— probably four—to which it would be centrally located.

In addressing the opportunities and challenges implied by this concept, attention was focused mostly on Dutch conditions both for agriculture and for nutritional needs. Consideration was given to plants that already grow well in the Netherlands or which are suitable for growing there and which contain potentially useful molecular structures. Potatoes, hemp and lupins meet these criteria. Equally, consideration was given to ingredients that could be obtained from biomass and to food products that could be made from these which could substitute for conventional food products. In particular, evaluations were made comparing the production of pasta from grains with production of a pasta substitute made from potato-derived ingredients and comparing cheese made from milk with the production of cheese from lupin-derived ingredients.

Lupin seeds contain around 18% oil and 40% protein. The remaining biomass waste can be used for clean energy production. Under Dutch agricultural conditions, the potential production level of lupin-seed cheese based on typical lupin seed yields and the average content of oil and protein (as mentioned above) is around 5 tonnes/ha/yr. This compares with 1 tonne/ha/year for conventionally produced dairy cheese. This represents a factor-5 improvement in the land intensity of cheese production, even before taking account of the imported fodder needs of dairy cows. Further environmental gains could be expected to come from the system of precise agricultural production and husbandry facilitated by the boom technology. Looking toward these other eco-efficiency indicators, cheese produced directly from biomass-derived ingredients causes factor-9 to factor-21 less environmental stress overall than cheese produced using animal converters. Figure 5.5 presents the relative normalised environmental burden scores for dairy cheese and three different types of lupin cheese. Results for the potato-derived pasta are also given, showing a factor-8 improvement relative to conventional pasta production methods.

From work during the definition phase, it was concluded that it would be possible, in principle, to close the nutrient (and possibly also the pesticide) loops through the envisaged systems of agricultural production as well as to fully use all the materials and energy content of produced biomass, diverting a much higher proportion of biomass to food and materials production than is the case today and using the remainder for clean energy production. The technological breakthroughs needed are especially in the fields of plant breeding and genetic engineering and of developing appropriate fermentation organisms for the bio-refineries. This is because the useful ingredients contained within plant biomass often have very similar molecular structures. Separation of different ingredients downstream by chemical or other means is very difficult and would impose a potentially high environmental burden. For an eco-efficient treatment, downstream processing has to be avoided. This puts the emphasis on upstream processes. In particular, it emphasises the need to develop plant varieties that have a large content of useful

Figure 5.5 **Total normalised environmental burden scores for two food products, pasta and cheese, produced traditionally and unconventionally (per kg of production)**

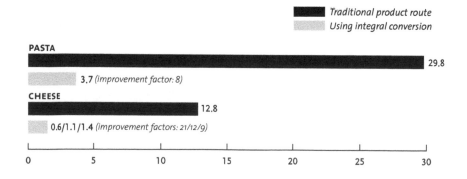

molecules which can be easily separated from other ingredients and the need to develop biological molecular separation techniques perhaps based on engineered microorganisms.

Follow-up work includes a Dutch contribution toward a project in Denmark that will see construction and operation of a pilot plant with a capacity to handle 10,000 tonnes of biomass per year.

HIGH-TECHNOLOGY-CONTROLLED CLOSED-CYCLE HORTICULTURE

As with integral crop conversion, the definition phase of the high-technology, controlled-environment research theme was aimed at more fully describing and evaluating the potential of ideas that had so far only been sketched out. These preliminary ideas had been developed by looking toward the major inefficiencies in current ways of meeting nutritional needs and by examining opportunities that today's nutritional systems overlook. In the case of controlled-environment horticulture, the major inefficiencies and lost opportunities arise because of the failure to make full use of sunlight, the major renewable resource needed for horticulture, and because of poor levels of control over the production process. There is a need to better manage the use of sunlight and to better control production processes. These two needs are strongly interrelated, since better control of plant growth in artificial and contained environments depends heavily on better control of sunlight and solar-derived energy.

The definition phase involved working with companies, consumers and research institutes. The companies spanned several different business domains. As well as enterprises involved in producing food, those with interests in promising new technological areas that might in the future play a role in food production in controlled conditions were also involved. The main concern was to raise levels of awareness among companies of the possibilities for increasing the eco-efficiency of food production in controlled environments. Consumer groups and their representatives were involved because of the wish to explore and include the social and cultural aspects of this form of food production, both in the technology designs and in the evaluation of the implications of pursuing this innovation trajectory. The research institutes, both private and public, were involved to help explore the technical aspects of individual technologies and technology clusters. These institutes made economic and environmental evaluations of different proposed technologies as well as suggested technology designs to be explored. Throughout this phase, the emphasis was on building a platform of support among this stakeholder network for some of the more promising ideas so that these might ultimately be taken up and carried forward without further STD support.

The concrete idea being worked up was that of a food production 'factory' located in or near to a city. One of the underlying principles is that as much food production as possible should take place close to the market so that transport and storage losses are reduced and so that supply can be more precisely matched with demand. Production close to consumption also means that organic wastes should be returned to production facilities to replace nutrients that leave in harvested foods. Re-use of organic waste is also the means of closing the CO_2 loop. Another key principle is that the food production facility is, as nearly as can be achieved, a closed system. The only recurrent inputs should be solar energy and make-up water to compensate for water removed from the unit in harvested products. A third principle is that of a completely controlled environment in which the delivery of CO_2, nutrients, water, light and heat should be regulated in respect both to plant needs and to the demand for foods expressed in local markets. By means of temperature, light and nutrient control, the growth and ripening of produce should enable harvests to be matched precisely to demands expressed in local markets. There should be no overproduction or storage loss.

The bottlenecks in realising this vision are primarily related to current technological limits on our abilities to use and manage solar energy. In principle, these bottlenecks could be overcome within the context of research and development on solar energy capture, steering, transformation and storage that is already under way, if the specific needs and opportunities presented by the nutrition sector are recognised and addressed. Recognising that the average receipt of solar radiation in the Netherlands of around 1,000 kWh/m^2/year is sufficient to meet all the light and heat requirements of growing plants, the technological problem is to manage this energy so that it is supplied to plants according to their needs rather than according to its natural availability. Improved conversion and storage of solar energy are aspects of the required technological solution. But an additional, related need is to be able to concentrate and steer solar energy so that it can be delivered precisely to a growing plant or to a specific part of a growing plant according to need. A set of related technological possibilities and opportunities were contemplated.

The first opportunity derives from the low use that plants make of solar radiation in photosynthetic processes. The overall efficiency of plant conversion of solar energy to plant tissue is, as indicated earlier, very small. It may be as low as 1%–2%. At issue is whether plants can be selectively bred or engineered to be more efficient in fixing solar energy and whether more of the received radiation can be converted to photosynthetically available radiation (PAR) within the 400–700 nanometre band. Photovoltaic cells that are transparent to PAR but which enable non-PAR energy to be converted to PAR could significantly increase the availability of useful radiation to plants. A specific technological need is for artificial lighting systems to provide plants with energy in wavelengths appropriate to their growing needs, using electricity derived from the non-PAR component of solar radiation. Moreover, since conversion involves transforming solar radiation to electricity and then back to light at a different wavelength, this process should enable energy to be moved from point of receipt to another location, which could even be in underground or multi-storey growing factories.

The second technological need is to enable the unused part of the solar radiation—the PAR fraction that passes through the photovoltaic cells but is not taken up by plants—to be used to produce heat. The surplus heat should then be stored for later use at night or during seasons when solar radiation is only poorly available naturally. Surplus heat could, in principle, be transferred to groundwater using heat pumps. The low exchange and flow rates of groundwater, together with the high capacity of water to store heat energy, makes groundwater a suitable medium for this purpose. It is also a readily available storage medium. Stored heat can then be returned as needed by reversing the heat pumps.

A third need is for solar energy that is received on rooftops and other surfaces where no plants are being grown to be captured and diverted to productive use. Photovoltaic cells can be used to accomplish this. However, owing to conversion losses, better use of the radiation could be made if the PAR content was directly transported or 'steered' to growing plants. This suggests that technologies are needed for concentrating solar energy—for example, using parabolic mirrors—and for transporting the concentrated sunlight into the enclosed growing environment. There are already optical fibres capable of performing this task, although these are far from optimised. Polymethylmetacrylate (PMMA), produced by Mitsubishi, is one such material. The distance that sunlight can be transported currently is around 200 metres. To prevent losses of light energy the fibre needs to be shielded with a coating that has a low refractive index at the relevant wavelengths. Preliminary work suggests that this can be a fluor-containing polymer coating. A range of other fibres and coatings suitable for performing this task need to be explored.

A fourth need is to close the nutrient and water cycles. In a tightly controlled production system, the need for nutrients and other additives should be significantly reduced because of better information being available on plant needs and because of better targeting of interventions within the production facility. Compensation, in the form of make-up materials, such as water, CO_2, macronutrients and micronutrients, is needed today only because of losses to the external environment. Today's glasshouses use 25 kg of water and 2.5 kg of CO_2 for each kilogram of produce. Only a small fraction of these amounts (only one-tenth in the case of the CO_2) leaves the glasshouse within the harvested produce; 90% is lost as fugitive emissions. In effect, then, the same technological breakthroughs for sensors and closed-loop production systems are needed here as in other projects within the nutrition sub-programme. In addition, in a tightly controlled urban production system there are also further opportunities for closing materials cycles because of the ease with which materials leaving the production facility in the form of produce can be recovered from households' organic wastes and returned to the facility. In principle, an almost complete recovery of nutrients should be possible in an urban setting.

A fifth need is for better harmonisation between production and consumption. Again, this is more easily achieved in an urban-based food production facility than in remote production systems because of the physical proximity of production to the market. Lower transport and storage losses and greater transparency of market conditions should allow production to be better oriented toward meeting exact demands. Information about local needs and preferences (tastes, lifestyles, etc.) should be better integrated than now into the production process so that only foods of a type, quality and quantity that exactly meet specified consumer requirements are produced. Because production in controlled facilities is intended to serve a small and localised community, feedback loops can be built into production processes so that the correct products are also produced just when needed. It is envisaged that information links between production units and retail outlets, such as food stores, would enable production and consumption to be dovetailed in terms of variety, quantity, quality and timing, while the technical means for achieving this will come through a higher control over growing conditions made possible by regulation of temperature and lighting conditions at the level of production units or even of individual plants. Production can then be focused exclusively on qualitatively acceptable foods and demonstrated demands.

Such a high-technology-controlled production system was evaluated within the project and found to offer opportunity for major gains in eco-efficiency and major reductions in environmental stress. Fossil energy needs may be reduced almost to zero. No recurrent fossil energy is needed for heating, lighting or to generate CO_2. In principle, CO_2 needs may be reduced by a factor of 8 and water use by a factor of 18 (Fig. 5.6). Since the material cycle is closed, contaminants are not lost to the external environment. Since this evaluation focuses on improvements in the efficiency of glasshouse production only, and not on potential gains made by better integrating demand and supply or by reducing transport and storage losses, the improvement potential on the supply side as a whole is anticipated to be greater still.

A very significant programme of research would be needed to break through the technological bottlenecks in plant physiology, solar energy management and closed-cycle production methods called for by such a vision. Some of these can be

Figure 5.6 **Inputs to glasshouse cultivation today per kilogram of product compared with high-technology closed-system production**

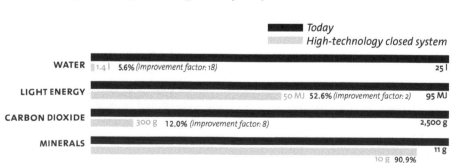

tackled in the Netherlands, but it was clear early on in the research that the scales of both the work involved and the potential benefits from it warrant joint-working within an international network. Carrying the ideas through to implementation would need a critical mass of interested parties and a division of labour. Within the definition phase, therefore, efforts were made to establish which of the necessary technological breakthroughs was already the subject of interest or work by research teams elsewhere and for what reasons. Clearly, the Dutch entry point for this research derives from the already well-established glasshouse industry and its significance both for the national economy and national nutrition. In the USA, NASA has a strong interest in producing food in controlled environments because of its space programmes. In Japan, institutions have been motivated to work on high-technology closed-loop food-production systems by the possibilities of developing systems that do not require a large amount of space. On behalf of Dutch actors in this field, STD has therefore explored interest in the possibilities for international collaboration. The outcome has seen the setting-up of an international network in which Dutch research teams will play a leading role. In addition, two Dutch companies, Cebeco Handelsraad and Nijseen Koelen, both commercial participants in the project, have committed to develop a pilot scheme.

OBSERVATIONS AND CONCLUSIONS

Research within the nutrition sub-programme demonstrates the different steps and phases of the STD approach more generally. In addition, the sub-programme exemplifies some of the more significant outcomes to be achieved along the way. These include one or more strategic visions of a sustainable future, a set of alternative innovation trajectories, a set of near-term innovation challenges, networks of innovators interested in pursuing the required R&D and platforms of support for promising ideas from stakeholders. Additional outcomes include a better awareness of the kinds of criteria to be used in designing and evaluating possible solutions.

The nutrition sub-programme clarified the advantages that could lie in achieving several different technological breakthroughs. Some can be considered to relate to bottlenecks that are of economy-wide significance for sustainable development. Finding solutions would open the door to progress on several fronts. Examples are photovoltaic technologies, biotechnologies and sensor technologies. These are critical for enabling progress in the field of nutrition such as achieving a switch from fossil to solar energy and making more efficient use of the solar resource (photovoltaics) or closing materials cycles (sensor technologies). But these same technologies are important also for sustainability in many fields of need, such as mobility, materials, lighting, heating, clothing and recreation. Equally, the work demonstrates that, although sustainability may depend on breakthroughs to overcome key technological bottlenecks, no single technological breakthrough of itself will prove decisive. Clusters of related technological breakthroughs that operate synergistically to support entirely new ways of meeting needs are likely to be needed in most cases. PAR-transparent photovoltaic cells, polymer-coated optical fibres, heat pumps, and sensors coupled to intelligent decision-support systems are all needed to realise food production in closed-cycle glasshouses. These technology clusters will, in their turn, depend for their effectiveness and acceptability on a related set of structural and cultural innovations. There is little point in producing foods in artificial settings from unusual ingredients or from genetically modified plants unless these are acceptable to consumers.

A final point is that some of the best opportunities to create sustainable futures in one sector or need field are dependent on breakthroughs in other sectors. Closing the nutrient cycle, for example, which is necessary for food systems to become sustainable, depends on breakthroughs in waste-water management. Unless the household waste-water stream is separated from other waste-water streams that contain hard-to-remove contaminants, such as heavy metals, contained nutrients cannot be recycled, nor can the nutrient cycle be closed. For this reason, sustainable nutrition depends on progress in waste-water handling, which in turn depends on actions such as changing patterns of use of heavy metals in industry which are, in part, beyond the direct control of water systems managers. In effect, all the various sectors and need fields are linked together as interlocking and interacting systems. As a consequence, progress has to be made systematically across all sectors if the sustainability potential of each individual sector is to be fully realised. Clearly, this will take time. It also calls for a strategic perspective in identifying critical linkages between sectors and the breakthrough innovations that will open up sustainability opportunities across them all.

Chapter 6

Constructive Technology Assessment: the case of novel protein foods

The previous chapter provided a broad overview of the work of the STD nutrition sub-programme. Against the backdrop of the high environmental burden of today's food system, five different innovation tracks were proposed that might allow nutritional needs to be met in the future at a substantially reduced environmental cost. Mostly these involve technological innovations on the supply side. Nonetheless, supply-side technological innovations have to be acceptable to consumers if they are to be effective in reducing environmental burdens. This link between the supply and demand sides is especially important in the case of nutrition, since, as consumers, we are all highly sensitive to issues surrounding the foods we eat. Cultural and religious factors are involved, as are traditions, customs and prejudices. There are also health and ethical factors. All these factors, and several others, have a bearing on the substitution and diffusion possibilities for new foods and foods produced in new ways. They create a good deal of uncertainty and risk for those capable of carrying forward an innovation track. There is little point in investing resources in a potential breakthrough technology unless there are strong grounds to believe that consumers will take up the new products and find the production methods acceptable.

The proposed innovation track for novel protein foods is a particular case in point. It engages extreme consumer sensitivities and is therefore a high-risk venture for supply-side innovators. Yet it also has the potential to offer substantial environmental and commercial benefits. The problem is that uncertainty and risk can kill innovation before it begins. They can prevent us from ever knowing whether a potential innovation track does, indeed, hold the promise that we think it could hold. We may never clarify what the potential costs and benefits of a technology could be or whether there are ways for society to manage a technology so as to maximise the benefits while minimising the costs and risks. Ironically, the innovation process itself offers the chance to learn by doing. By starting the innovation process, we can move along the learning curve and, by asking and answering the right questions, reduce the uncertainties and risks of further development. At the same time, the innovation process can be used to gain greater understanding about stakeholders' concerns, to factor these concerns into technology designs and to build platforms of understanding and support from stakeholders for agreed innovation pathways.

In this chapter we review some issues involved in the development of novel protein foods and the information needs to which these give rise. Our main aim is to describe how a particular methodological approach, Constructive Technology Assessment (CTA), was used in the project to tackle the problem just described. CTA was used to initiate a dialogue between potential innovators of novel protein foods and stakeholders. The objectives of that dialogue were to explore the interests and concerns of all parties, to review available information about novel protein foods relevant to those concerns and to identify significant information gaps. Through an inclusive, interactive and iterative process that alternated between workshops aimed at identifying questions and formal research aimed at providing answers, it was ultimately possible to structure a consensus around a long-term diffusion scenario for the introduction of novel protein foods and to develop short-term research activities based around a set of promising novel protein food ingredients. Support was built both for an R&D programme encompassing specific source–technology–product combinations and a detailed market introduction plan. In this chapter, we describe and evaluate the experience with CTA in the novel protein food project and present the findings of the project about novel protein foods.

ISSUES IN THE DEVELOPMENT OF NOVEL PROTEIN FOODS

Two projects within the STD work on nutrition looked at the technological possibilities for finding environmentally more benign substitute products and production routes for meat and dairy protein. One was the project on integral crop conversion that looked toward ways of recovering edible molecules, such as oils, fats and vitamins, from non-edible biomass, and making substitutes to conventional protein-rich products from these, such as the lupin cheese described in Chapter 5. The second project, the focus of this chapter, looked for new ways of obtaining protein-rich materials from 'alternative sources' and converting these into food ingredients that could provide an attractive and acceptable meat substitute. The kinds of 'alternative sources' in question could include genetically modified plants, algae, fungi, bacteria and other protein-rich materials. These two projects were identified during the orientation phase of work on the nutrition sector and were developed further during the definition phase when the basic ideas were also fleshed out and the opportunities and barriers to these innovation tracks were explored.

The motivation for both projects comes from the environmental inefficiencies and unsustainabilities introduced into food production by the use of animal converters, the background to which has already been described. The strong roles of meat in the Dutch diet and in the economy, especially as represented by the production, consumption and trade in pork, impose a high demand for eco-capacity, much of which is met by eco-capacity 'imports'. Using animals to convert grain feedstock to animal protein involves the loss of 80%–90% of the contained nutritional value of the feedstock, which introduces a substantial eco-inefficiency. It also leads to a large and often concentrated production of organic waste. Moreover, analysis of the eco-efficiency improvement potential in the traditional meat production chain shows this to be limited. A 10%–20% reduction in acidify-ing and eutrophying emissions from animal husbandry is estimated to be possible over the next ten years through manure treatment. Thereafter, further reductions will become increasingly more difficult and expensive to achieve and, because of the energy requirement of manure treatment, will come only at expense of shifting the environmental burden to other areas such as climatic change and fossil fuel depletion (Heidemij cited in de Haan *et al.* 1997).

One possible way of reducing this environmental burden and dependency is to develop substitute products that might fulfil the function of meat but which are produced through more eco-efficient routes (Box 6.1). However, even if eco-efficient technological routes for meeting protein needs could be demonstrated, sustain-ability benefits will only be reaped if these foods are acceptable to consumers. Even a 20% take-up of novel protein foods, with a resource productivity improvement over meat of a factor 20 at the product level, would reduce the environmental stress of protein production at a systems level by only 19%. This clearly illustrates how much environmental performance improvement at the systems level is dependent on market penetration levels (diffusion) and not only on technical performance at the level of the production process or product. This is especially critical in the early stages of adoption when market penetration rates may be so low that even eco-efficiency jumps at the product level of factor 30, 40 or 50 may have little or no impact at the system level (Fig. 6.1).

This raises a number of serious questions over the links between technological innovation and cultural and structural innovation. Many of the opportunities for technological eco-efficiency improvement at the level of production processes and products lie in fields such as genetic engineering, which engage the sensitivities and concerns of consumers, especially over health and safety implications. There are also established cultural values and habits to be considered. While foods fulfil nutritional needs, this is not their only function. The foods we eat provide satis-

THERE ARE ALREADY PRODUCTS ON THE MARKET THAT PROVIDE NON-MEAT PROTEIN: for example, tofu. Some of these are marketed as stand-alone products and others as meat substitutes in 'meat-like' forms, such as vegetarian sausages. Today, most sales are to highly specialised consumer groups—such as vegetarians and vegans—and the products have not achieved rates of general market penetration, either in the Netherlands or in comparable Western societies.

The production process for novel proteins—of any sort—has three steps. The first is to grow the protein substrate, such as soy, wheat or bacteria. The next step is to extract and refine the protein. The third step is to texturise the proteins (Fig. 6.2). In principle, two different types of end-product can be realised. The protein can be left as an unstructured, single-cell protein (SCP) or made into a texturised vegetable protein (TVP) with a composite structure for use as a meat analogue. TVPs can also be laminated to produce specific meat analogues, such as bacon substitutes.

These production processes are much more eco-efficient than meat production processes on a product-equivalent basis (i.e. per kilogram of final protein produced). Figure 6.3 describes the environmental impact of meat production and Figure 6.4 gives comparable data for the production of a meat analogue based on SCP. In both cases, the data is broken down by stage in the production chain. The major environmental burden of SCP production is caused by energy use. But even this level of energy use represents a factor-5 reduction relative to meat production. On land use and biocide dispersion, SCPs show factor-18 reductions versus meat. On all other concerns and indicators, the reduction is greater than a factor of 100. Table 6.1 extends the analysis to include other meat alternatives.

Despite the significantly lower eco-capacity requirements during production of novel protein foods compared with meat and dairy products, their impact on the overall sustainability of the protein sector depends also on diffusion rates. Taken together, SCPs, TVPs and TVP-laminates have less than 1% of the total protein market today.

Box 6.1 **The relative eco-efficiency of meat and novel protein foods**

faction through their flavours, aromas and textures. They also say something about us. Foods are used both to confer and confirm social standing. Meat is widely considered to be a luxury item and a status good. Eating food together plays a major role in social relations among people, families and groups. Important relationships and family occasions are marked by eating important foods. The cultural perception of meat and the associations it entails are deeply entrenched in beliefs and habits that are self-perpetuating. All in all, the concerns of consumers over unconventional foods together with the strong entrenchment of conventional foods in eating norms and habits constitute significant barriers to dietary change.[1]

If novel protein foods are to contribute to any meaningful reduction in environmental stress, they clearly need to achieve a substantial market penetration. This

1   There have been several negative experiences to date linked with the failure of some early novel protein foods and, more recently, a developing controversy in some European countries (such as the UK) over the use of genetically modified organisms in the human food chain.

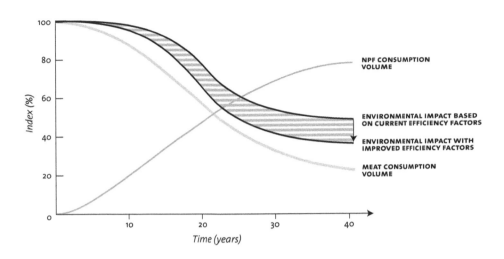

Figure 6.1 **The overall reduction in environmental impact will directly correlate to the volume of meat being substituted.**

Figure 6.2 **The production process for novel proteins has three steps.**

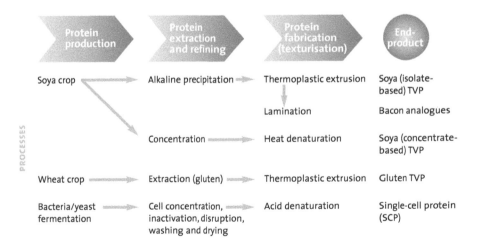

Environmental concern	Crop growing/processing	Transport of feed components	Feed processing	Animal breeding	Animal fattening	Animal slaughter and meat processing	Other transport	Meat preparation	Absolute value
Depletion of:									
▷ Fossil fuels*	52%	17%	4%	15%		2%	5%	4%	39.8 MJ
▷ Copper resources			100%						0.06 g
▷ Zinc resources			100%						0.2 g
Use of land*	76%			7%	17%				14.6 m²
Climate change				40%	60%				2.9×10^{-7} Ceq
Acidification		11%		34%	51%		3%		1.7×10^{-6} Aeq
Eutrophication	9%			33%	58%				2.7×10^{-8} Eeq
Dispersion:									
▷ Biocides	100%								0.2 g
▷ Copper	1%			64%	34%				0.3 g
▷ Zinc	2%			12%	86%				0.2 g

4.4 kg feed — Relative impact of the different steps in the value chain on the environment — *1 kg pork*

* Based on van Eerdt *et al.* 1993

Figure 6.3 **Meat production contributes to a wide variety of environmental concerns.**

Single-cell proteins

300 g molasses → Crop growing/processing → Bacterial fermentation* → Centrifuge separation and drying → Acid denaturation → Transport and preparation → 1 kg meat analogue (15% protein)

Environmental concern	Crop growing/processing	Bacterial fermentation*	Centrifuge separation and drying	Acid denaturation	Transport and preparation	Total	Reduction factor versus meat
Depletion of:							
▽ Fossil fuels (MJ)	1.4	0.6	1.7	~0	3.7†	7.4	5
▽ Copper resources (g)						0	>100
▽ Zinc resources (g)						0	>100
Use of land (m²)	0.8					0.8	18
Climate change (10^{-7} Ceq)		0.0015				0.0015	>100
Acidification (10^{-6} Aeq)						0	>100
Eutrophication (10^{-8} Eeq)	~0					0	>100
Dispersion:							
▽ Biocides (g)	~0					~0	18
▽ Copper (g)						0	>100
▽ Zinc (g)						0	>100

* Mixing, aeration, sterilisation, cooking

† Assuming same transportation and preparation as for meat

Figure 6.4 **On a weight-for-weight basis, NPFs can significantly reduce most environmental concerns.**

ENVIRONMENTAL CONCERN		SCP	MYCO-PROTEINS	TVP	PLANT PROTEINS	DAIRY PROTEINS	FISH PROTEINS	LABORATORY/TISSUE CULTURE PROTEINS
					Reduction factors relative to meat			
Depletion of:								
▷ Fossil fuels		5	3	7	9	n/a	n/a	L
▷ Copper resources		>100	>100	>100	>100	>100	>100	>100
▷ Zinc resources		>100	>100	>100	>100	>100	>100	>100
Use of land		18	6	6	100	n/a	n/a	>100
Climate change		>100	>100	>100	>100	n/a	>100	>100
Acidification		>100	>100	>100	>100	n/a	>100	>100
Eutrophication		>100	>100	>100	>100	n/a	>100	>100
Dispersion:								
▷ Biocides		18	6	6	100	n/a	n/a	>100
▷ Copper		>100	>100	>100	>100	>100	>100	>100
▷ Zinc		>100	>100	>100	>100	>100	>100	>100
Overall environmental efficiency		M–H	M	H	H	n/a	n/a	L–M

H = high; M = medium; L = low

Table 6.1 **On a weight-for-weight basis, NPFs can significantly reduce most environmental concerns.**

raises the question of whether such a level of market penetration could be achieved and, if so, how. But it also raises concern about the impact such a level of market success would have on the established meat industry. A high level of market penetration by novel protein foods could negatively effect jobs, income and export earnings in the established meat sector and its upstream supply industry. Today's meat-protein chain contributes strongly to the Dutch economy. Around 1.75 million tonnes of pork is produced annually in the Netherlands today; 60% of this production is exported. The meat-producing sector directly provides 140,000 jobs, contributes US$6 billion to national income and delivers a net US$3.5 billion to the balance of payments. Where would successful innovation of novel protein foods leave the established meat industry?

Concerns such as these over the likelihood of achieving consumer acceptance of novel protein foods and over the potential negative economic and social effects of meat substitution on a large, internationally competitive and already-established industry have contributed to an unwillingness on the part of both private and public actors to take this innovation path forward. This is an innovation field where success depends on substantial investment. The high risks and uncertainties mean that investment is not forthcoming soon enough or in sufficient amounts in the prevailing market context. Nevertheless, in principle, it should be possible to produce improved novel protein foods and this route offers the potential of substantial long-term environmental benefits. Moreover, there are some opportunities and positive signs in recent trends and developments in a number of areas that may tip the scales in favour of revisiting this technology field if an acceptable way can be found to reduce the risks and uncertainties of investing in the necessary R&D.

PROMISING TRENDS AND DEVELOPMENTS

One of the trends that makes for an optimistic outlook on novel protein foods is the value migration in consumer food choices already described in Box 5.1. A set of newly emerging values is beginning to influence these choices and the concerns reflected in these values are both spreading and strengthening. There is a growing concern over the health and safety implications of all foods, but especially of meat. In part this is related to increased scientific understanding of the importance of nutrition, the nutritional status of different foods and links between nutrition and human health. Equally, there is growing scientific understanding of the aetiology of human diseases and of the role of food production processes and specific foods as risk factors. As well as concern over the role played by a high consumption of meat and dairy products in conditions that are common and well known, such as cardiovascular disease, new concerns have been raised by the incidence of specific disease transmission from animals to humans through the human food chain and by epidemiological and aetiological studies that have increasingly implicated meat production methods and meat consumption *per se* in the aetiology of prevalent human diseases. Bovine spongiform encephalopathy (BSE), bird-flu, *E. coli* and salmonella are recent examples. The reputation and image of meat have been damaged. Opportunities for novel protein foods may be found as consumers increasingly seek acceptable alternatives.

These health concerns are matched by concerns over the ethical implications of food choices. Consumers' awareness of the environmental, animal welfare and equity implications of intensive meat production methods and of meat consumption continues to grow steeply. Recent consumer research in several countries shows that there is already a large pent-up demand, so far largely unmet, for foods produced in ways consistent with rapidly emerging 'green' and 'ethical' values held by a growing number of consumers. A survey conducted by a large supermarket chain in the UK toward the end of 1997, asking consumers about which categories of product they would most like to see stocked, revealed that organically produced foods and foods produced under conditions humane to animals were the most wanted commodity options. Moreover, related research shows that many consumers are prepared to pay a price premium for foods produced to standards consistent with these emerging values. At present, the bottlenecks are mostly on the supply side. Production costs are too high—and, therefore, prices to consumers are still too high—for most consumers to be able to translate their actual food preferences into consumption choices. Nevertheless, the pent-up demand and the ethical values on which it is predicated might provide openings for future product differentiation and market segmentation.

Backing up these trends in value migration, there have also been some positive recent developments in the alternative proteins market. Recently, the take-up rates for meat alternatives as stand-alone ingredients have improved. In part, this reflects a shift towards ready-made meals that traditionally have been based on mechanically recovered and minced meat but are increasingly being based on meat alternatives. However, the major impetus has come from the launch of 'Quorn', a new novel protein food developed and marketed by Astra-Zeneca. Quorn can be considered to belong to a first-generation of new novel protein foods on account of its substantial improvement over earlier meat analogues and its broad positioning on the protein food market. Quorn is superior to these forerunners in texture, appearance, the biological value and availability of its protein content, its capacity to retain added flavours and its lack of undesirable off-notes and side-effects (Table 6.2). It has broken out from the restrictive market segment of vegetarians and

Table 6.2 **The commercial attractiveness of Quorn is higher than that of meat analogues, but still insufficient to expect large-scale adoption.**

Commercial criteria	WEIGHT FACTOR	TEXTURISED SINGLE-CELL PROTEIN	TEXTURISED VEGETABLE PROTEIN	LAMINATED TOPS	TEXTURISED MYCOPROTEINS (QUORN)
Sensory experience	20%	L	L	L	M
Nutritional and health value	10%	L–M	M	M	M–H
Socioeconomic acceptability	15%	L–M	M	M	M
Conformity to lifestyle trends	15%	L–M	L–M	M	M
Attractiveness to producers	40%	L	L–M	L–M	M
Overall score	100%	L	L–M	L–M	M

H = high; M = medium; L = low

vegans traditionally catered for by novel protein foods to find appeal among the wider public. Quorn provides evidence that novel protein foods can be acceptable to consumers *if* appropriate quality criteria are met.

Quorn also represents a significant technological breakthrough over earlier meat substitutes and offers guidance on where further technological breakthroughs are needed to produce even better novel protein products. Quorn is a texturised mycoprotein, based on a mould substrate grown on molasses. The mould substrate is produced by continuous fermentation over a period of 20 weeks. The protein extraction step is achieved by cell concentration, heat inactivation and separation using centrifuges and filters. The final stage of texturisation to produce the end-product is achieved by heat setting and chill denaturation. All aspects of the production process represent fundamental steps forward in the technology of novel protein food production. The mould fermentation process required break-through R&D on the fermentation process itself, the selection of mould strains and the development of safety and toxicity screening protocols. In addition, incre-mental R&D was needed in fermenter design, aseptic separation, automatic pro-cess control and co-extrusion (Fig. 6.5).

Further technological improvements at all stages in the production process of novel protein foods are possible. Several new protein sources could be considered, including natural or improved vegetable protein sources, de-novo synthesis and, in the longer term, animal tissue culture. Natural protein sources include non-edible biomass, algae, seaweed and fungi. Improvement methods to increase pro-tein yields or improve quality include classical strain improvement, genetic

Figure 6.5 **An existing first-generation NPF, Quorn, represents a technological breakthrough.**

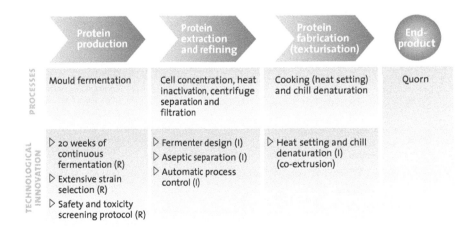

R = radical R&D; I = incremental R&D

engineering and protein engineering technologies. De-novo synthesis involves manufacturing proteins by the enzymatic coupling of amino acids. Animal tissue culture involves growing animal tissues in laboratory-like conditions from sample animal cells in the same way as replacement human tissues and organs for transplant surgery are beginning to be produced in experimental conditions today. Finally, several radically improved texturisation and fabrication technologies are conceivable based on different forms of protein engineering such as spinning, stretching and cross-linking (Fig. 6.6).

By combining different protein sources with different production process technologies, it is possible to construct at least 24 categories of protein foods. Figure 6.7 charts the possibility set and shows the position within the potential product portfolio of conventional protein product groups such as meat and dairy foods and existing novel protein foods. These occupy only four of the available 24 sites in the potential product portfolio. The remaining 20 product categories can be considered to represent different combinations of a production technology and a potential protein source. A sequence of successively higher-quality novel protein products could be realised by applying successively higher-level technologies to each protein source. The sequence of technological improvements starts with the application of existing fabrication technologies to new protein substrates. It proceeds through the application of radically improved fabrication technologies:

Figure 6.6 **Further technological improvements at all stages of the production process can create higher-quality and more varied NPFs.**

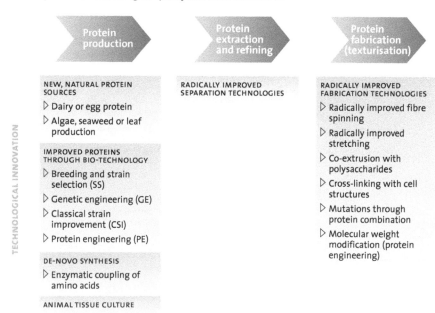

TECHNOLOGY

	PROTEIN SOURCE					
	Single-cell organisms (bacteria, yeast)	Multi-cell organisms (moulds)	Vegetables (soya, gluten)	Plants (algae, seaweed)	Animals (chicken, dairy cows, meat)	Laboratory
Existing product	Texturised single-cell proteins	Texturised mycoproteins	Texturised vegetable proteins		Meat	
New product						
a. New protein substrates texturised through existing fabrication				Texturised plant proteins	Texturised dairy and egg proteins; Texturised fish proteins	
b. Existing or new protein substrates texturised through radically improved fabrication	Superior texturised single-cell proteins	Superior texturised mycoproteins	Superior texturised vegetable proteins	Superior texturised plant proteins	Superior texturised dairy and egg proteins; Superior texturised fish proteins	
c. Strain-selected proteins texturised through radically improved fabrication	Strain-selected superior texturised SCPs	Strain-selected superior texturised mycoproteins	Strain-selected superior texturised vegetable proteins	Strain-selected superior texturised plant proteins		
d. Genetically engineered/CSI proteins texturised through radically improved fabrication	GE/CSI superior texturised SCPs	GE/CSI superior texturised mycoproteins	GE/CSI superior texturised vegetable proteins	GE/CSI superior texturised plant proteins		
e. Non-texturised proteins						Plant tissue culture; Animal tissue culture; De-novo protein

Figure 6.7 **Twenty categories of new NPFs can be obtained when combining various technologies.**

first to new protein substrates, then to strain-selected protein substrates and finally to genetically engineered protein substrates. In a final level of technology improvement, the production of texturised protein is envisaged, such as by animal tissue culture and the production of de-novo proteins.

By building on the successes and learning experiences of earlier source–technology combinations, the risks and uncertainties in the innovation process can be reduced. Moreover, the development costs for the near-term possibilities (the source–technology combinations that could be reached within the next five years) are relatively low compared with the more ambitious long-term options. The costs of development for products within the higher-level and longer-term categories are high. However, on the basis that these options are developed on the back of learning achieved by producers and consumers with earlier novel protein foods, the chances of successful innovation in these higher-level product categories are high. An assessment of the costs and risks is given in Table 6.3. New protein substrates texturised through existing methods could be developed within two years at an estimated annual cost of US$1 million (c. 2 million Dutch guilders). For this, existing technology requires only to be incrementally improved. At the extreme end of the ambition spectrum, tissue culture and de-novo protein production would require up to 30 years of R&D and an annual budget of US$10 million. The chance of technological success with these routes is estimated at around 30%.

In principle, then, the prospects for novel protein foods are good because there are opportunities to differentiate novel protein foods from each other and also from meat. Moreover, because they can be more easily customised than meat, novel protein foods could surpass meat in terms of sensory properties (aroma, texture, taste, etc.), nutritional and health properties (increased fibre, lower fat, lower cholesterol, etc.), and conformity with lifestyle trends (increased convenience, greater versatility of use, better keeping properties). The routes for novel protein food production involve a large number of technologies, most of which are at an early stage of development. Today we are far from the natural limits of these technologies. The substantial number of different combinations among protein substrates and processing technologies offers a wide opportunity set of potential product types and characteristics. In principle, waves of successive product types could be produced over coming decades, each taking longer and costing more to develop but each building on the success of earlier products and achieving ever-higher quality and rates of market penetration. Because a product sequence can be envisaged, we can be reasonably sure that this is not a technological blind alley, but an alternative development trajectory with substantial potential (Fig. 6.8).

Nonetheless, novel protein foods still need to be substantially improved if they are to replace a significant share of meat in total protein consumption. Large-scale substitution of meat is needed for there to be meaningful reductions in the environmental burden of human protein nutrition, but this can be achieved only if novel protein foods have the quality to be positioned as stand-alone products.

NPF category	TECHNOLOGY STATUS	R&D TYPE	R&D ANNUAL BUDGET (DUTCH GUILDERS)	R&D TIME-FRAME (YEARS)	CHANCE OF TECHNOLOGICAL SUCCESS
New protein substrates texturised through existing fabrication	M	I	2	2	100%
Existing or new protein substrates texturised through radically improved fabrication	G–M	I–R	5	5	100%
Strain-selected proteins texturised through radically improved fabrication	G	R	5	10	95%
Genetically engineered/CSI proteins texturised through radically improved fabrication— SCP, mycoproteins	E–G	R	10	8	90%
Genetically engineered/CSI proteins texturised through radically improved fabrication— vegetable, plant	E	F	10	15	90%
Plant and animal tissue culture; de-novo protein	E	F	20	30	30%

E = embryonic; G = growth; M = mature; F = fundamental R&D; R = radical R&D; I = incremental R&D

Table 6.3 **New NPFs can be developed within 20–30 years with a high chance of technological success.**

This will come about only if novel protein foods can achieve a broad position through a range of new products. The sticking point is the risk and uncertainty involved to innovators. The acceptability of novel protein foods to consumers remains the central concern. This is an excellent example of a situation where there are real sensitivities involved in technology development and where the costs and risks of making wrong choices are high.

This makes the development of novel protein foods a particularly interesting and challenging case study. For the STD programme, the issue was less to provide unquestioning support for a novel protein food development programme than to establish an inclusive process by which the value of such a programme could be assessed and its design influenced by stakeholders. This can best be achieved if all parties work together. The design and evaluation criteria for both the technologies

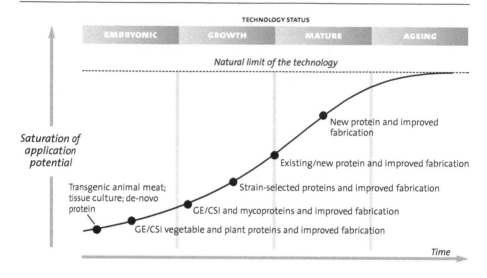

Figure 6.8 **The R&D programme for new NPFs will require largely embryonic and growth technologies.**

and the diffusion process need to be worked out jointly if the full range of potential concerns of all interested parties are to be represented, which is necessary if obstacles to successful innovation are to be pre-empted and avoided. In situations such as this, there is need to support the innovation process by widening the information basis of decision-making. Earlier variants of technology assessment methodology are poor supports, but newer methods, such as Constructive Technology Assessment, can be helpful.

CONSTRUCTIVE TECHNOLOGY ASSESSMENT

There have been several forms of technology assessment methodology that have been used to provide information to decision-makers about innovation processes. The earliest forms of technology assessment sought only to evaluate the probable innovation impacts of technologies that had already been designed. In later variants of the methodology, attempts were made to influence the technology design process by introducing additional design criteria that analysts considered should be met in order to avoid adverse consequences during innovation and diffusion. Although this represents a more proactive approach to technology development, it maintained a separation between the technology developers and stakeholders. Stakeholders are neither actively consulted nor engaged in the process directly. Rather, the analyst decides which criteria to include in the design process and how important these are. This presupposes that all the relevant criteria are discernible without consulting stakeholders directly and that the positions and perspectives of stakeholders are fixed both in regard to the technology and in regard to other stakeholders. Neither of these assumptions is likely to be valid.[2] Attitudes to a new technology are likely to depend on how the technology is designed, why it is being introduced and how the innovation process is managed.

Constructive Technology Assessment (CTA) is a relatively new variant of technology assessment methodology that has been pioneered in the Netherlands. The major difference between it and earlier methods is that, in CTA, the analyst plays a neutral role. Values come from stakeholders directly and not from the analyst. The role of the analyst is therefore different in CTA; it is to facilitate a discussion among innovators and stakeholders through which relevant design and evaluation criteria as well as the positions of stakeholders can both emerge and be formed. This has several advantages. The criteria come directly from stakeholders themselves and can be used from the outset to influence the innovation path. The set of relevant criteria is likely to be more wide-ranging and inclusive than that presupposed by analysts. Criteria are also likely to be revealed as fungible and conditional. The discussion process can be used to explain the context of the innovation to stakeholders, which might influence attitudes and positions. The process could also reveal information gaps and uncertainties which, if clarified, might lead to shifts in stakeholders' positions. Finally, the CTA process provides opportunity for a creative dialogue between innovators and stakeholders in a constructive and 'power-neutral' context.

CTA is therefore a more useful and versatile tool than earlier technology assessment methods. It can play a wider set of roles and perform more functions. CTA is particularly useful in the context of complex and unstructured problems, which are typical of sustainable technology development.[3] Moving toward sustainability implies taking decisions about new and unproven technologies with uncertain costs, benefits and risks. New technologies are, by definition, characterised by uncertainty. This means that new technologies need to be introduced via

---

2   In practice, the values of stakeholders may not be clear either to the analyst or even to stakeholders themselves. Positions taken by stakeholders will depend on the information that is available to them on the context of the technology development, including its implications for other stakeholders and for society in general. Uncertainties and information gaps can cloud issues, but can be removed through research.

3   Unstructured problems are those where there is a high degree of uncertainty about the relevant facts and where there is disagreement about relevant values. Structured problems, in contrast, are ones where the facts are clear and where there is a good deal of consensus in the relevant value systems of different stakeholders. Uncertainty and dissent are typical of problems facing those involved in sustainable technology development. CTA can provide structure by forging alternative problem definitions.

a process that involves open and frank discussion about concerns, risks and contexts, which seeks to integrate concerns into technology designs and which builds consensus around agreed courses of action to reduce uncertainty. CTA is useful because it contextualises innovation management as a participatory social process of learning, consensus building and risk/responsibility sharing. By definition, this dictates that CTA is an inclusive, interactive and iterative process involving information gathering, information exchange and decision-making.

CTA is *inclusive*. In CTA, a key objective is to ascertain *all* stakeholders' concerns either directly from them or from their representatives. Accepting that the values and positions of different stakeholders are likely to be in conflict and that significant uncertainties about the future are likely to cloud perceptions and positions, a specific objective of CTA is to develop a possible synthesis. This does not necessarily imply a compromise but rather a search for entirely new positions to which the many different parties can all subscribe. This may often be related to the need to reduce uncertainty by gathering information that can help clarify everyone's position. CTA seeks to give all stakeholders indirect influence over the forms that technology and innovation take even if they have no direct influence. It does this by seeking to *integrate* revealed values and preferences into decision processes as relevant decision criteria. For this, it is necessary for CTA to be an *iterative* process. The process tends to move between rounds of discussion and rounds of research. The research is driven by the questions raised during the workshops and is designed to feed new information back into subsequent workshops. Progress depends on amassing information about technological possibilities and the consequences of progressing these. The typical products of CTA are *agreed* action plans representing a synthesis position for short-term actions toward *agreed* longer-term objectives.[4]

THE USE OF CTA
IN THE NPF PROJECT

Within the STD programme, it was thought worthwhile to revisit the issue of whether novel protein foods have a role to play in meeting future nutritional needs. This led to the decision to stimulate a review process to consider the issue, which, in the event of the parties involved agreeing to pursue this innovation track, would also increase the probability of a successful innovation outcome. CTA was used to

---

4   A CTA can be considered successful if, as a result of repeated circles of argumentation among participants: all were equally capable of helping to generate the analysis; positions and views shifted during the analysis toward a new synthesis; participants gained a greater understanding of each others' views; the process inspired creativity; the process equipped the parties for action; action plans were agreed; and learning experiences were transferred by participants back to the groups they represent.

provide a bridge between the preliminary evaluation work made entirely by STD and any work undertaken autonomously by independent innovation agents in the future to develop and market specific novel protein foods. The CTA was therefore intended to deliver three tangible outcomes. There should be a decision by food and biotechnology companies over whether to give a new impetus to the development of novel protein foods. This decision should be backed by the creation of a research consortium with sufficient resources to carry out the necessary R&D without further STD help. Such a decision should be influenced by the CTA process and reflect its second tangible outcome. This should be an agreement on promising novel protein food options, a timetable for the market introduction of these and the content of a supporting programme of R&D. The third tangible outcome should be an action plan for activities to support and smooth the innovation process, involving actions by government agencies and leadership groups within society.

These tangible outcomes depend on the CTA process achieving success in terms of consensus building and information gathering. The workshops and supporting research activities are therefore the key organising elements of any CTA. The main issues to decide are who will participate in the process and how will supporting research activities be organised. Preliminary decisions about these are taken by the facilitator to set the process in motion, but these are reviewed and amended by participants during the CTA. Invitations to participate in the novel protein food CTA were issued to enterprises, organisations and institutions involved in food research, food production and food marketing, including industrial, technological and marketing consultancies. Representatives of the main consumer groups and environmental groups were invited. Also, representatives of various government departments and agencies were involved. Government has a special role because it is the agent representing the public interest. It must decide where the public interest lies and be concerned over the structural and distributional consequences of innovation, even when that innovation is considered to be in the overall public interest. Government is also an important actor because it can support new technologies directly and indirectly and can smooth the process of economic restructuring.

Some specific questions addressed by the CTA are listed in Box 6.2. Table 6.4 lists the research activities set up to answer these and shows how these were organised into six themes. Several different research institutions and university departments within the Netherlands had responsibilities for the different themes (Fig. 6.9). To be successful, the CTA needed to accomplish a shift from discussing novel protein foods in general to discussing specific source–technology combinations that could be explored and evaluated by participants. A key objective was to screen a wide set of possible novel protein foods to arrive at a shortlist of promising technologies for detailed scrutiny. The process can be seen as one of profiling and evaluating specific novel protein foods in the context of innovation scenarios capable of achieving targeted levels of market penetration and environmental improvement. By backcasting from targeted systems-level eco-efficiency improvements, it was

▷ Can technologies be developed that meet revealed criteria? Which technologies offer the best chances? What are their development costs? How long is the time to market?

▷ How much will novel protein foods cost to produce? How big is the potential market? How might the market be developed? What are the potential returns on investment?

▷ What are the necessary supporting measures that government and other actors need to take? Will these actions be taken?

▷ What eco-efficiency gains at the system level could be achieved using novel protein foods? By when? Are the gains sufficient to justify public support for the innovation track?

▷ What public policies will be needed if this innovation track is followed?

Box 6.2 **Specific questions addressed by the NPF CTA**

Table 6.4 **Broad information needs of the NPF CTA**

RESEARCH AREA	OBJECTIVE
1. Technology and product development	▷ Detail required R&D programmes for selected NPFs
2. Consumer	▷ Identify desired stand-alone properties of NPFs
	▷ Determine acceptability of NPFs to consumers
	▷ Forecast consumption patterns and NPFs' market volume
3. Environment	▷ Detail environmental efficiency of selected NPFs
	▷ Estimate overall environmental impact under different substitution scenarios
4. Business economics	▷ Calculate financial returns under different substitution scenarios
5. Structural implications	▷ Determine impact on Dutch economy and politics under different substitution scenarios
6. Communications	▷ Determine how best to inform actors and stakeholders about NPFs

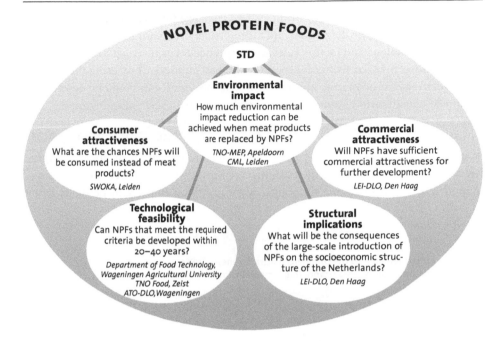

Figure 6.9 **Research responsibility in the NPF study**

possible to develop marketing plans for novel protein foods capable of meeting these targets. The implications for technological, cultural and structural innovations and for restructuring could then be explored and decisions made on what short-run actions to take in order to meet targets.

The kinds of question that the CTA needed to address, the sensitivities of this research area and the need to help broker agreements in regard to the setting-up of new research consortia implied that the project should be led by a senior manager. A well-respected leader with a strong background of industrial experience was appointed to facilitate the CTA process. He was supported by the STD programme office and a research team drawn from specialised research institutions in the Netherlands.

THE RESULTS OF CTA IN THE NPF PROJECT

One of the first tasks was to identify criteria relevant for the selection, design and evaluation process and rank these according to their relative importance. By way of example, Table 6.5 lists some of the criteria that are important for the commercial success of novel protein foods, including factors relevant to consumers when making food choices and factors relevant for producers as revealed during the CTA.

Which novel protein technologies offer the best chances of meeting the performance criteria? In principle, hundreds of novel protein foods could be developed. Using the classification system for potential products described in Figure 6.7, the different possibilities were screened using revealed criteria to search first for promising protein substrates and then for protein ingredients that could be made from them. The most promising potential protein sources are peas, lucerne, lupin, the fungus *Fusarium* and the cyanobacterium *Spirulina*. Three ingredients—termed 'protex', 'fibrex' and 'fungopy'—were identified that could be made from

Table 6.5 **Criteria relevant to the commercial attractiveness of NPFs and their relative importance (high, medium, low)**

DEMAND OR SUPPLY	DIMENSION	PROPERTY	H	M	L
Demand side	Sensory experience	Flavour	x		
		Texture	x		
		Appearance	x		
		Absence of off-notes		x	
		After-taste		x	
		Satiety			x
Demand side	Nutrition/health value	Protein content/quality	x		
		Digestibility/tolerance	x		
		Safety	x		
		Other nutrients			x
Demand side	Socioeconomic acceptability	Price/price:quality	x		
		Public acceptance	x		
		Status/luxury good		x	
Demand side	Conformity with lifestyle trends	Quality image	x		
		Fresh/healthy/natural		x	
		Convenience		x	
		Animal-friendly			x
		Environment-friendly			x
Supply side	Attractiveness to producers	Production cost	x		
		Return on investment	x		
		Growth opportunities	x		
		Time to market		x	
		Price pressure		x	
		Investment level		x	
		Framework conditions			x

these. All score well on factors linked to commercial attractiveness (to producers and consumers), reduction in environmental strain and minimisation of adverse structural economic effects. Each ingredient is representative of a different structural form. Protex has a structure comparable to minced meat; fibrex has a fibrous structure; fungopy has a solid structure similar to tempreh. These three ingredients can be derived in several different ways using different source–technology combinations. Table 6.6 shows seven promising routes to these novel protein ingredients from the different sources just listed.

How well do these seven source–technology options meet the eco-efficiency and nutritional prerequisites for this innovation track? An examination of the entire production chain for pork and the seven novel protein food ingredients was made using life-cycle assessment. For both pork and the novel protein food options, the highest environmental strain occurs early in the chain where biomass is fixed and bioconversion is completed. The processing of raw materials contributes less than 10% to the respective indices in either case. Figure 6.10 compares the environmental index of the selected novel protein food ingredients relative to pork. The first column shows the calculated environmental index for pork using 1995 production technology. The factors that make the largest contribution are acidification, nutrification and aquatic eco-toxicity. The data set out in Figure 6.10 reveals that each of the selected novel protein food ingredients can be produced in ways 5–32 times more eco-efficient than pork. However, the seven novel protein foods do not all offer the same comparative eco-efficiency gains over pork production. The best performance is expected from protex derived from *Spirulina*. This is because this production route does not require a previously cultivated vegetable substrate as the *Spirulina* bacteria fixes its own CO_2.

Figure 6.10 also shows where improvements in the eco-efficiency of producing the novel protein food options could be made and how big the improvements could

Table 6.6 **Seven high-potential NPF options**

INGREDIENT	PROTEIN
Protex: an ingredient resembling minced meat in structure that can be made from bacteria, yeasts and plants	1. *Spirulina* (cyanobacterium) 2. Pea 3. Genetically modified pea 4. Lucerne
Fibrex: a fibrous ingredient made by continuous fermentation of fungi	5. *Fusarium* (fungus)
Fungopy: an ingredient produced by fermenting plants with fungi	6. Pea with *Rhizopus* (fungus) 7. Genetically modified lupin with *Rhizopus* (fungus)

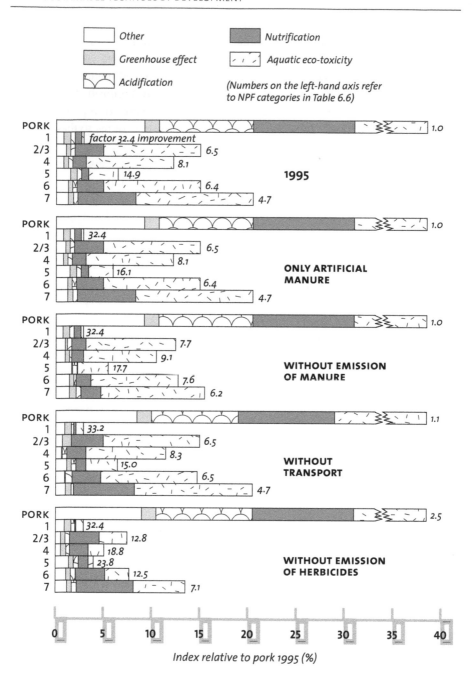

Figure 6.10 **NPF options: current environmental impact expressed as an environmental index relative to pork and the indices following improvement**

be. For the options based on a vegetable substrate rather than on *Spirulina*, the major environmental stresses are nutrification and aquatic toxification associated with substrate production. Any reduction of herbicide and fertiliser use in agriculture, which is what would be needed to reduce the eco-capacity requirement of pork production, would also improve the environmental performance of novel protein food production.

Further environmental gains in the protein sector therefore depend on research to reduce the environmental burden of producing protein substrates, especially work on improving the pest-resistance of crops and on environmentally friendly pest control methods. There seem to be no technological obstacles to achieving sustainable systems for producing protein substrates. Expansion of ecological farming could be an important step in the short term. In the longer term, the development of biological pesticides and strains more resistant to pests offer the best prospects. By 2035, such improvements could make it possible for novel protein foods to be produced 80 times more eco-efficiently than pork is produced today. By then, we can expect some improvements to be made also in the eco-efficiency of pork production. However, the same improvement in substrate production would make pork production only 1.4 times more eco-friendly than it is today. This means that on a weight-for-weight protein-content basis, novel protein foods would be 60 times more eco-efficient to produce than meat by 2035.

Are these novel protein foods as nutritious as meat? Meat supplies the body with essential amino acids and a range of specific vitamins and minerals. For each of the seven novel protein foods studied, Figure 6.11 shows the content of these main vitamins and minerals relative to pork. The data needs to be interpreted with care, because the ability of the body to absorb the different components of food can depend on the co-presence of several nutrients and the physical form of specific components. It can also be inhibited by the presence of phytate and lectins. The form of iron that occurs in meat, for example, is more readily absorbed than the inorganic iron contained in vegetable foods. Similarly, Vitamin B12 has been shown to have lower absorption rates from vegetable sources than from animal sources. While only small quantities of B vitamins are present in the novel protein options studied, the concentration of these could be increased if this is considered important nutritionally. More generally, however, research is needed into ways of improving the absorption rates of vitamins and minerals from vegetable-derived proteins rather than of reinforcing the vitamin and mineral contents.

The next step in the CTA process was to develop a market plan for the introduction of novel protein foods on the assumption that foods as good as meat or better than meat can be developed. Workshop sessions suggested that consumers would be reluctant to forego meat altogether, especially in the 'pure' form of steaks, chops and roasts. The marketing opportunity for novel protein foods is therefore most likely to lie in the processed market segment as a substitute for mechanically recovered meat, as a replacement for sausage and minced meat in particular. An

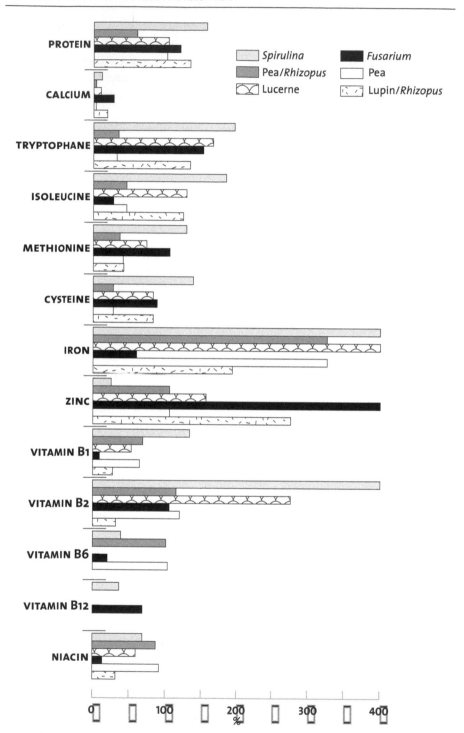

Figure 6.11 **Composition of NPF raw materials in percentages relative to pork (= 100%) on a dry-substance basis**

analysis of trends in meat consumption shows that the processed segment of the market is growing. It currently accounts for 45% of the total protein market. If present trends continue over the next 40 years, the processed segment would account for 75% of the total protein market. This is mostly due to the switch to convenience foods and ready-made meals. During the CTA, the emphasis was therefore placed on the substitution possibilities for novel proteins in the processed sector. Different possible futures were explored on the basis of achieving different substitution levels. The likelihood of being able to achieve these substitution levels was related to marketing information about the probabilities of different types of people and households being willing to try novel protein foods and about trends in eating behaviours. Younger people, for example, are more likely to experiment with different foods and are more flexible in their behaviours than are older people.

Several future visions were developed around different rates of eco-efficiency improvement at the systems level, different rates of eco-efficiency improvement of novel protein foods at the product level and different rates of novel protein food substitution for meat. From these future visions, different scenarios were set out representing different possible tracks toward a sustainable future. A base scenario was constructed using a conservative (average factor-20) eco-efficiency performance for novel protein foods at the product level and a goal of halving environmental strain at the systems level by the year 2035. This base scenario would require that novel protein foods take just over half of the processed segment of the market by the year 2035. This is the equivalent of a 40% share of the overall protein market. Based on inferences such as these, timetables were developed from now until the target date for product performance improvement and for the diffusion of novel protein foods. In effect, these set out what actions will need to be taken and what progress will need to be made in order to keep on track. In this way, long-term targets were translated into a timetable for the achievement of near-term goals.

A near-term goal consistent with the long-term target of the base scenario is for novel protein foods to achieve a 5% market share within ten years and for novel protein food production, by then, to be five times more eco-efficient than pork production (Table 6.7). The table suggests how daily patterns of protein consumption would need to change between now and 2035 through meat substitution. The 5% target by 2005 could be reached if all persons in households in the Netherlands headed by 20–40-year-olds replace meat once a week with a novel protein food product. If, by 2035, all 20–65-year-olds and their children replace the meat content of three main meals each week, meat consumption would fall by around 30%. Any further reduction must be sought either in the substitution of meat in a fourth meal during the week or in an increase in the tendency to consume several smaller meals at shorter intervals throughout the day. CTA participants considered the substitution target for 2005 realistic, but substantial measures would be needed to achieve the target for 2035.

YEAR	MEAT (g/person/day)	NPF (g/person/day)
1995	117	0
2005	111	6
2015	94	23
2025	76	41
2035	70	47

Table 6.7 **Substitution trajectory of NPF protein for meat: 1995–2035**

How attractive is this baseline scenario as a commercial proposition? Major factors in answering this question are the costs of novel protein food production relative to meat protein production and the size of the potential market. Production costs depend on the source–technology combinations and the envisaged production volumes, but costs for all seven source–technology combinations are substantially lower than for meat production. Research by LEI-DLO indicates that per-kilogram production costs will be less than US$0.5 for the vegetable options and less than US$1 for the microbial options. There is therefore ample scope for using novel protein food ingredients to prepare low-cost end-products, which makes them potentially attractive to the ready-made meal industry. Moreover, the low production cost and the enormous size of the protein market—700,000 tonnes per year in the Netherlands alone—mean that even small rates of market penetration can deliver substantial returns to producers. A 5% market penetration level by 2005 would already make novel protein foods an attractive business development proposition. The attraction is increased if novel protein producers sell their products directly to consumers by making and marketing ready-prepared meals themselves. The overall size of the protein market is, of course, vast. A 40% share of the protein market by 2035 would mean an annual production of 300,000 tonnes of novel protein foods for the Dutch market alone. There is also a huge potential export market.

Would the environmental benefits warrant the innovation? Assuming that the basic substitution scenario outlined in Table 6.7 was to be achieved, the total strain on the environment arising from protein nutrition would be reduced by 55% by the year 2035 (Table 6.8). This calculation is based on the assumption that the reduction factor of 20 is achieved for novel protein foods on average and that, at the same time, a reduction factor of 1.4 will have been achieved in the environmental strain of pork production. A greater reduction becomes possible if ecological production methods in agriculture become more commonplace. A higher average eco-efficiency performance improvement at the product level would enable the

YEAR	PERCENTAGE OF MARKET (MEAT)	REDUCTION FACTOR (MEAT)	PERCENTAGE OF MARKET (NPF)	REDUCTION FACTOR (NPF)	PERCENTAGE REDUCTION IN TOTAL STRAIN
1995	100	1.0	0	5	0
2005	95	1.2	5	5	20
2015	80	1.3	20	5	35
2025	65	1.3	35	10	47
2035	60	1.4	40	20	55

Table 6.8 **Total reduction in environmental strain per kilogram protein consumed relative to 1995**

same level of stress reduction to be achieved at a systems level with a lower rate of meat substitution.

What would be the wider socioeconomic effects of the level of meat substitution envisaged in the base scenario? An important finding of the supporting CTA research activity was that there will be some significant structural changes in the meat sector in coming decades, even assuming autonomous, business-as-usual, development. Autonomous development will lead to a smaller meat sector by the year 2035. World trade liberalisation, tighter environmental policies and reform of subsidies are all likely to contribute to a gradual relocation of pig farming within Europe toward grain-producing areas. By 2035 the Netherlands could anyhow expect, therefore, to maintain fewer livestock. Employment within the meat-producing sector is likely to be reduced by up to 50% as a consequence of such autonomous changes. The overall sector income is also likely to fall in real terms by 20%, although the overall contribution of the sector to the Dutch balance of payments—the net value of all meat and feedstock trade—is likely to remain the same as now.

If large-scale substitution of meat by novel protein foods was superimposed on these autonomous changes, there would be further job losses in the meat sector, estimated at around a further 28,000 person-years. However, job losses in the meat sector would be partially offset by the creation of new jobs in novel protein food production, estimated at around 13,000 new person-years. These will mostly come in the agricultural sector in the production of vegetable protein bases. The annual production of 750,000 tonnes of novel protein foods—300,000 tonnes for the domestic market and 450,000 tonnes for the export market—would require around 270,000 hectares of arable land. On balance, this would still lead to a net loss of 10% of the workforce in the combined protein sector. The contribution of the protein sector to balance of payments would also fall by one-third. The contribution expected to be made by novel protein food exports will be less than the anticipated fall in the contribution from meat exports net of changes in the value of trade in feedstocks (Table 6.9). In short, while there are substantial environmental benefits arising from this scenario, there are significant socioeconomic consequences of restructuring which would need to be handled well through complementary social and political measures.

OUTCOMES AND CONCLUSIONS

To what extent can we consider the interactive approach taken in this project to have been successful? The available technologies for novel protein food production are not yet sufficiently attractive to consumers to secure a significant reduction in meat consumption, but they have the potential to become much more attractive and, in the process, could reduce environmental stress significantly. However, this is not an easy or straightforward innovation field. As well as technological obstacles, there is a host of other obstacles to overcome which are rather less tractable. In this field, the technology developers are especially dependent on co-evolutionary cultural and structural innovations over which they have no direct influence, which makes the innovation track a risky one. A signal of support from stakeholder groups within society is important for the active technology developers and for government, both of whom can feel more secure in investing resources knowing that the research line is backed by important opinion leaders in society.

Table 6.9 **Socioeconomic consequences of an NPF substitution scenario**

INDICATOR	SCENARIO	1995	2005	2015	2025	2035
Market share	Autonomous	100	100	100	100	100
(%)	NPF substitution	0	5	20	35	40
Meat consumption	Autonomous	89	89	89	89	89
(kg/capita)	NPF substitution	89	85	71	58	53
NPF production	Autonomous	0	0	0	0	0
('000 tonnes)	NPF substitution	0	94	375	656	750
Contribution to GNP	Autonomous	12.1	8.9	8.5	9.4	10.4
(billion Dutch guilders)	NPF substitution	12.1	8.7	7.7	7.8	8.2
Net contribution to BoP	Autonomous	6.9	7.0	6.0	6.6	7.2
(billion Dutch guilders)	NPF substitution	6.9	6.1	4.7	4.5	4.7
Employment	Autonomous	140	94	76	75	71
('000 person-years)	NPF substitution	140	91	68	61	56

The use of CTA is important in such contexts because there is a need for all stakeholders to agree on a course of action. Ideally, the situation calls for a way forward that is jointly developed by all the parties whose support is needed for the track to be successful. Through continuing CTA interaction, those active in technological, cultural and structural innovation can come together to explore the problem and to work out a new joint approach. The CTA process is as important as its tangible outcomes, since the process itself provides the opportunity to explore the reasons for the innovation track and to appreciate each others' positions. Within the CTA, there is the possibility that stakeholders' positions will change as information is made available and that a platform of support can be built around a joint action plan. A jointly developed action plan that integrates stakeholders' concerns will provide a much stronger basis for successful innovation than one worked out in isolation and may lead to a different assessment of whether the innovation track is even worth pursuing.

STD's novel protein food project positioned the problem of future protein nutrition in the context of sustainable development, which allowed new synthesis positions to be explored and developed. It was possible to find common interests among stakeholders who had initially come to the project from very different backgrounds and positions. The concerns of stakeholders were used to highlight uncertainties and information gaps that could be addressed through research. From this it was possible to establish a constructive discussion and to find new ways forward. Through the CTA process, new protein chains, representing alternatives to today's dominant meat protein chain, were described and evaluated. Through the CTA process, it was possible to build consensus around a long-term diffusion scenario for the introduction of novel protein foods and to develop short-term research activities based around a set of promising novel protein food ingredients. Support was built for an R&D programme encompassing specific source–technology–product combinations, a set of R&D priorities in novel protein food development, a detailed market introduction plan and a programme of supporting actions designed to smooth the innovation path.

The CTA also generated substantial amounts of detailed information about the environmental stresses of the meat chain and the possibilities of reducing these using novel protein food. This information is needed to lobby for financial support for novel protein foods, for outreach programmes of education and for designing policies to support plans for novel protein food introduction. The CTA has alerted government to the likely scale of the structural changes that would arise in the case of a significant substitution of novel protein foods for meat and has given the political system time to create acceptance for such changes and to develop supporting policies.

By translating consumer demand into explicit product quality standards, the CTA was able to identify areas for research on technological bottlenecks (Table 6.10). Research plans were set out to 2010 for short-term research in the fields of

RESEARCH FIELD	RESEARCH NEED	TERM
Sensory science	▷ The characterisation and modification of the structural properties of proteins	Short
	▷ The interaction of proteins with taste and aroma components	
	▷ The applications for NPF ingredients within consumer end-products	
Reducing eco-impact	▷ Environmentally friendly pest control and sustainable strain improvement	Short
Nutritional value and health	▷ Functional components and components that enhance general health	Short
Commercial production	▷ Production upscaling	Mid

Table 6.10 **Priority research fields for NPFs**

improving the sensory, nutritional and environmental performance of novel protein foods. The CTA identified the need, also, for mid-term research in the field of production upscaling. This technological research was made part of a wider programme of R&D activities designed to address other, non-technological, obstacles to this innovation track, including research on relevant consumer behaviour, possible resistance to change from parties likely to be adversely affected by it (such as meat producers), and concerns over the political and other consequences of structural change.

As a consequence of the CTA, the entire R&D programme has been fully adopted by one of the Netherlands' top technology institutes with support from both public and private sources. Supporting work on novel protein foods is being carried out also at research institutes in Delft. These research activities are receiving sponsorship from several companies involved in the project, such as Unilever, Gist-brocades and Boekos.

A significant programme of communications and educational support has been started, using the project results to inform and influence the debate on cultural innovation. The major activities are programmes of information and education aimed at the general public (using social networks) and at young people (using formal institutions of education). In addition, a programme of work has been established on policy instruments, including market instruments and measures to alleviate any adverse consequences of economic restructuring. All three innovation lines—culture, structure and technology—have, thus, been stimulated.

Chapter 7

LCA as a steering tool: the case of the municipal water system

Today's municipal water 'system' has been developed over a long period during which the sustainability of the system was not a major concern. The water 'system' includes all the technological, organisational, economic and behavioural aspects of municipal water management in respect to a 'chain' which begins with the extraction of water from the environment and ends with its return there. A tacit assumption of the development and management of the system is that water is abundant and that eco-capacity is not a constraint on the system. The physical system infrastructure and the management and use of the water system have been established against the backdrop of this critical assumption. But the assumption is no longer valid and the system is already showing signs of serious problems. Stress is placed on the environment at all stages of the water chain. This chapter explores the design and implementation of an alternative, more sustainable, municipal water system. It follows the approach taken by the STD water project and focuses, especially, on its use of life-cycle assessment (LCA) as a steering tool to guide the search for a more sustainable system. The chapter draws heavily on the work of Hugo Meijer and Jaap van Leeuwen of STD's Water Project (van Leeuwen and Meijer 1996; Meijer and van Leeuwen 1997) and a set of expert reports written by Professors van den Akker, Terpstra, Wiggers and Lettinga (van den Akker *et al.* 1997).

Through the use of LCA it was found that the design and operation of today's water system is incompatible with sustainability and that no amount of fine-tuning of today's system could bring about sustainability improvements of the level sought by the STD programme. Achieving major improvements in sustainability depends on defining a new strategic approach and on restructuring the water system using new criteria and insights. Equally, by using LCA it was found that no one measure can deliver sustainability. Rather, restructuring depends on making many changes to the design and operation of the system and achieving 'system-wide' synergies among these. Improvements from combined measures are much greater than the sum of individual environmental improvements achievable by stand-alone measures. Shifting toward a more sustainable system is less a matter of technological innovation than of organisational and behavioural innovation. A problem is that responsibilities for the system are divided and no one authority has the operational control needed to implement a new system alone. Improving the system therefore depends on the relevant parties agreeing on strategic targets and committing to take concerted action. Even then, there are limits to what can be achieved by changes on the supply side. Some of the most important actions need to be taken on the demand side by system users in households, businesses and industry. LCA proves to be a powerful aid in developing action plans and building platforms of support for the necessary changes. One tangible result has been the signing of a pact of co-operation by representatives of the water authorities, water companies and district councils in the region of Almelo aimed at implementing a sustainable water system by 2040.

STRATEGIC PROBLEM ORIENTATION

Today's water system is characterised by the chain that begins with water 'winning' and ends with the return of used water back to the environment. At the beginning of the municipal water chain, water is extracted from rivers and groundwater stocks. It is purified to drinking-water standard at centralised facilities and then distributed through a network of pipelines to households, offices, hospitals, service facilities and industries where it is used for drinking, cooking, bathing and showering, washing clothes, watering gardens, cleaning, and flushing toilets. Water is used, additionally, as an input to many production processes in industry and to perform various functions that exploit one or another of its very special physical and chemical properties. Water is used in heating and cooling systems, in hydraulic systems, as a solvent, to hold materials in suspension and as a medium for transporting wastes. Some municipal water becomes embodied in products or is lost dissipatively but most, once used, enters the sewerage system together with any wastes and impurities picked up during use. It becomes waste-water and is pumped to centralised sewage treatment facilities. In the general sewerage system, which is used as a convenient drainage system, it is joined by storm-water from roads and from the urban environment. At the sewage plant, waste-water is treated aerobically to remove organic and inorganic impurities and 'freshened' to increase the dissolved oxygen content before being returned to the aquatic ecosystem. The procedure yields contaminated sewage sludge, which is mostly landfilled or incinerated.

The design of the physical system infrastructure and these patterns of system use and operation reflect the influence of several factors. One is the perception that water is neither a scarce resource nor one whose use imposes a high burden on eco-capacity. Another is that water has been perceived, in economic terms, to be in the category of a pure public good. Universal provision of water services is considered to be in the general public interest because of the economic and health implications of non-provision. Provision requires statutory powers because of the need to build networks of pipelines and sewers across private land. Also, there are difficulties in establishing and implementing appropriate pricing arrangements. Charging for water services on the basis of actual use has been, until recently, either difficult or impossible. As a result of such factors, the water industry has been located, institutionally, in the public sector. The responsibilities of the water industry for ensuring a supply and delivery of clean water to users, for safe-guarding the environment and public health and for protecting urban areas from flooding are viewed as public service obligations. These are currently discharged by the operation of water production and storage facilities, purification plants, water delivery pipelines, the sewerage network and waste-water treatment plants.

Owing largely to the assumptions underpinning these institutional arrangements, the water industry has long been dominated by a supply-side approach to balanc-ing the demand and supply for water services. There has been little attempt to manage and reduce demand. On the contrary, supply has been demand-driven while demand has grown as a consequence of charging arrangements, such as flat rates and subsi-dised prices which often encourage higher usage. Because clean water is available 'on tap' at low or zero marginal cost to the user, there is no disincentive to overusing water. From the purely economic perspective of efficient resource allocation, levels of potable water use are therefore excessive. Parallel arguments apply to waste-water. Households and industry routinely use drains and sewers as a convenient means for disposing of general wastes. Common abuses include the use of the sewerage system by households, industry and business to dispose of toxic liquid wastes, such as used oil, paint and solvent. The activities of system users are neither easily moni-tored nor directly controllable by the water industry. Nonetheless, these have a profound effect on the possibilities for the water industry to meet its obligations and on the economic and environmental costs of doing so. Moreover, because it is easy for waste producers to offload the cost of waste treatment onto society as a whole in this way, the system establishes a disincentive to waste reduction at source.

Water is one of our most basic resources, but neither the production of clean water nor the treatment of waste-water is currently sustainable. In the Netherlands there are five main problem areas:

▷ Groundwater extraction is one of the main causes of the desiccation of farmland and natural ecosystems.

▷ Energy is consumed at every stage in the water chain.

▷ Shortcomings in waste-water treatment methods mean that surface water becomes polluted by heavy metals and substances that consume oxygen.

▷ Contaminants leak into the soil via rainwater that has fallen on roofs and roads outside built-up areas and via breaks in sewers.

▷ In the production of potable water and in the treatment of waste-water, contaminated sludge is produced for which there is no useful application.

What features of the physical system design contribute most to these sources of unsustainability? First, the physical system is highly centralised. In order to reap economies of scale, water production and treatment facilities tend to be few in number and large in size. This means long transport distances for potable water and waste-water, which adds to the energy requirements of operating the system. Second, the waste-water system is only partial. The cost of connecting remote and isolated users to the sewerage system is high. To hold costs down, such users often remain unconnected and their waste-water goes back to the environment without treatment in a centralised facility. Similarly, the cost of catering for peak demand, such as occurs during severe storms, is high. Sewerage systems are therefore designed to handle only base-loads. During peak periods, when the capacity of sewage treatment facilities is overreached, raw sewage is allowed to overspill directly into river systems. Third, the materials used to build the physical infrastructure of the water system include water-soluble heavy metals such as copper, lead and zinc. These are slow-release sources of toxic contaminants.

The pattern of use of the system also contributes to unsustainability. The most important factor is the high volume of clean and waste-water passing through the system, since this leads to a high demand for water at the beginning of the chain, which contributes to desiccation, and a high demand for water treatment at the end of the chain. However, the additional significance of the high volume of waste-water is that this leads to dilution of the organic waste content, which reduces the efficiency of the water treatment process. The dilution also affects the efficiency with which heavy metals can be removed from the waste-water stream as these bind with the organic matter. A concentrated waste-stream is easier to clean up. The low concentration of organic matter in the waste-water forces the use of aerobic treatment processes, which produce more sewage sludge than anaerobic processes. Moreover, because the sewage sludge produced by the treatment processes contains heavy metals, it cannot be used as a fertiliser. Large quantities of unusable sludge are produced. The nutrients contained within the sludge represent a loss of potentially useful material. This poses a double environmental stress, since the sludge becomes a toxic solid waste that must be incinerated or landfilled, while its disposal in either of these ways prohibits closure of the nutrient cycle.[1]

---

1 There are still further environmental implications, since nutrients lost to agriculture through the contamination of sewage sludge must be made up by using artificial fertilisers, whose manufacture and use is resource-intensive.

The major reason for the high volume of water passing through the system is that only two qualities of water are recognised: potable water and waste-water. Potable water is used for every purpose whether or not water of drinking quality is actually needed to fulfil the function. Potable water is used for flushing toilets, washing clothes, cleaning cars and watering gardens just as it is for drinking or showering. Similarly, the sewage and waste-water treatment system is used as a general system for all waste-water, including both household waste-water and rainfall run-off from roads and built-up areas. The use of the sewerage system for removing rainfall adds to the volume of waste-water to be treated at the sewage plant, further reduces the concentration of organic matter in the waste-water and introduces additional heavy metal contamination from roads and roofs (Box 7.1). It also means that local rainfall is lost as a potentially useful source of 'grey' water for low-grade applications such as flushing toilets or watering gardens.

A life-cycle analysis was used to quantify the extent of the environmental stresses imposed by today's water system. The starting point was to specify criteria on which the environmental implications of the water system should be judged. Operational indicators were developed for five of the six sources of unsustainability outlined above. Desiccation was not operational within the LCAs since this is a local problem that depends on site-specific aspects of the draw on groundwater stocks. The five aspatial themes on which the water system was evaluated were its draw on non-renewable materials, draw on fossil energy, release of contamination

Box 7.1 **Sources of heavy metals that enter the waste-water stream**

DATA COMPILED BY THE DUTCH WATER AGENCY, RIZA, USING ITS 'PROMISE' MODEL (RIZA 1996/1997), shows that more than 1,000 tonnes of heavy metals are transported annually within the Dutch waste-water stream (Table 7.1). This includes mercury, cadmium, chrome, copper, nickel, zinc and lead. Some metals come from sources within the infrastructure of the water system itself. Zinc, copper and lead have been used routinely as materials for making gutters, pipes and solders. Lead flashing was once commonly used on roofs. Soluble metal ions within the materials of the physical infrastructure are dissolved by the continuous flow of water. In total, the annual contribution of sources from within the physical infrastructure of the water system is estimated to be 663 tonnes. In mass terms, the single most significant contaminant from within the system is zinc with almost 450 tonnes annually. Lead (113 tonnes) and copper (100 tonnes) are other significant contributors.

In addition, a further 345 tonnes of heavy metals comes from both point and diffuse sources outside the water system. Examples of important industrial point sources are plating works, tanneries and print shops (for chrome), photography workshops (for silver) and dentists' surgeries (for mercury). An important diffuse source is the transport system. Lead, zinc, cadmium and other heavy metals emanating from fuels, tyres, vehicles or road materials are washed into sewers from roof and road surfaces. Zinc is the most significant material in mass terms, with an additional 170 tonnes entering the water system annually from outside. Copper (42 tonnes), lead (22 tonnes), chrome (20 tonnes) and nickel (20 tonnes) are also important. A further 72 tonnes of heavy metals within the waste-water stream are from unknown sources.

	Hg	Cd	Cr	Cu	Ni	Zn	Pb	Total
Sources within the water industry infrastructure				99.7		449.7	112.9	663
Sources outside the water industry infrastructure	0.46	1.00	20.1	41.8	19.6	169.6	21.5	345
Unknown sources	0.22	1.26	26.6	0.4	24.4	19.0		72
Total	0.68	2.26	46.7	141.9	44.0	638.3	134.4	1,008

Table 7.1 **Annual heavy metal content of the waste-water stream in the Netherlands (× 1,000 kg); based on model data for 1993**

Source: PROMISE Model (RIZA 1996/1997)

to the aquatic environment, release of contamination to soils and generation of solid waste. In respect to each, the absolute stresses imposed by the water system were compared to all other stresses of similar type in the Netherlands to give a feel for the relative contribution of the water system to national unsustainability. By normalising the impact scores in this way, the significance of findings are easier to interpret. Especially, it becomes easier to define where improvement efforts should be concentrated.

The LCA of the present situation (Fig. 7.1) shows the environmental impact in respect to these five dimensions of sustainability.[2] Based on the normalisation routine that translates absolute stresses into percentage scores, today's water system accounts for just less than 1% of the total Dutch annual draw on non-renewable materials, slightly more than 1% of all Dutch fossil energy use,[3] almost 13% of the organic and inorganic pollutant contamination of the aquatic ecosystem and almost 8% of total toxic solid waste generation. Soil contamination is not significant in the current situation. These results highlight that the major sources of unsustainability in the current water system are its contributions to aquatic eco-toxicity and toxic solid waste production.

In turn, the high rates of aquatic toxicity and toxic solid waste production reflect inefficiencies in water treatment that are intrinsic to today's water system. Water treatment is not 100% effective. The effective removal rate for the organic content of waste-water at the treatment facility is 90%–95%. For the heavy metal content,

2   The current situation makes demands on eco-capacity in respect to four of the five themes. The fifth—soil contamination—is not a significant problem at present, but could become so if the system were to be restructured along lines considered and evaluated later in this chapter. Soil contamination is therefore included in Figure 7.1 to enable comparison of LCA findings across the project as a whole.

3   A breakdown shows that most of the energy is used for pumping water around the system or for mechanical oxygenation of water at treatment plants.

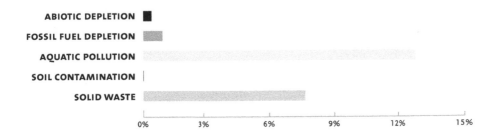

<figure>
Figure 7.1 **Environmental effects of the water chain**

Source: TNO-MEP 1997
</figure>

the rate is 70%–75%. In effect, some 5%–10% of the organic matter and 25%–30% of the heavy metal content of waste-water is not removed at the treatment facility and is discharged back to the aquatic environment in the 'cleaned' water-stream. For the Netherlands, the consequence is a discharge of 43,500 tonnes of organic and inorganic pollution from sewage plants into rivers. A further 6,500 tonnes of pollutant enters rivers because the treatment system is only partial: 5,000 tonnes of pollution escapes during storm episodes when flows of rainfall and sewage overreach the handling capacity of treatment facilities; a further 1,500 tonnes of pollution is generated by users not connected to the system. Taking these together, the average effectiveness for the waste-water treatment system as a whole is 57% for heavy metals and 72%–82% for organic matter (Fig. 7.2).

In fact, the situation is even worse than this, since even material removed from waste-water during treatment has still to be disposed of safely. It is important to track what happens to the contaminated sewage sludge. The current fate of sludge and recent trends in this are shown in Table 7.2. As a consequence of more stringent environmental regulations, an increasing share of sludge cannot be used in agriculture or for composting, which used to be important end-uses for sludge. Initially, sludge diverted from agricultural use was sent for landfilling. However, owing to the high water content, direct landfilling is giving way to incineration followed by landfilling of the produced slag and ash. This reduces the volume of the solid wastes by 70%. However, the advantage of compaction is offset by the disadvantage that heavy metals can be lost to the atmosphere during incineration of sludge or to groundwater and soils by leaching from incineration of waste products.[4] The loss of heavy metals to the environment from the municipal water system is, therefore, likely to be greater than the 43% direct loss reported above.

4   Slag can be used as an aggregate for road construction purposes, for example in stabilisation layers. But it is necessary to look closely at the leaching characteristics. Fly ash contains high concentrations of heavy metals and is usually classified as a hazardous material and landfilled under rigorous conditions.

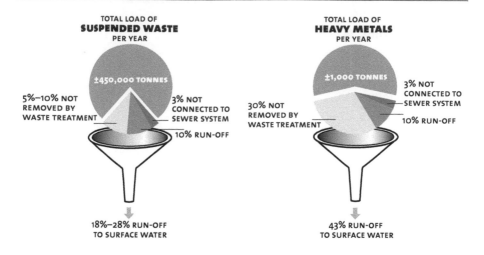

Figure 7.2 **Average effectiveness of waste-water treatment**

The current water system is expensive and elaborate. But perhaps most surprising is that the system is unsustainable in spite of all the resources devoted to it. Despite the efforts devoted to water treatment at sewage plants, upwards of 50,000 tonnes of pollutants enter surface aquatic ecosystems annually from the municipal water system, including almost 500 tonnes of heavy metal. The system also produces 3.2 million tonnes of unusable solid sewage sludge.

According to the working hypothesis of the STD programme, the aim should be to reduce these problems by a factor of 20 in a sustainable water system. A further indicative life-cycle analysis was therefore made based on the assumption that measures are taken so that all waste-water is collected and channelled to treatment facilities where cleaning is carried out until the level of contamination of the

Table 7.2 **Trends in the disposal and valorisation of sewage sludge**

Source: RIVM 1991

END FATE OF SLUDGE	1984	1992	1994
Agriculture	32%	22%	5%
Composting	31%	21%	24%
Landfill	33%	54%	50%
Incineration	4%	3%	21%

returned water-stream is low enough not to pose a threat to the aquatic environment. Such a near zero-emission system would represent the outcome of pursuing water treatment in today's water system to its ultimate end-point. It would require that the sewerage system is extended to embrace all water users and to meet all demands, including peak demands. The economic costs of implementing and operating such a water system would be prohibitively high. However, an indicative LCA of such a system also reveals some alarming environmental consequences. Any achieved reductions in aquatic eco-toxicity would be offset by a substantial increase in energy use. Energy use would be more than double today's level and even more unusable sewage sludge would be produced. The reasons for this can be traced to contradictions inherent in the design of today's water system as a large-scale realisation of what is effectively an end-of-pipe approach to environmental protection.

Today's approach represents an attempt to clean up after contamination has been introduced into the municipal water system rather than an attempt to prevent contamination from entering in the first place. By allowing organic and inorganic wastes to come together, it is preordained that sewage sludge cannot be used as a fertiliser. All contaminants removed during water treatment will become solid wastes. In effect, the contradictions inherent in the system ensure that waste-water treatment can only be concerned with transferring pollution between media. The higher the throughput of water and the greater the dilution and mixing of wastes, the higher the costs of effecting this transfer. In effect, there is no possible sustainable solution to be found in an approach to environmental protection within the water system based on end-of-pipe treatment. No amount of fine-tuning or extension within the prevailing model will enable it to become sustainable. We must look, instead, for a more efficient and differentiated solution. A crucial turning point in the investigation was the realisation that such a closed-system approach could not be sustainable.

SKETCHING A SUSTAINABLE FUTURE

How might a sustainable water system of the future look, say, two or three generations from now? One of the key features of the STD approach is the whole concept of looking ahead into the far future. The programme focuses not on current deficiencies and problems but on future needs and objectives. A conjectural sketch of a sustainable water chain of the year 2040 outlines what those needs might be and what the possibilities are for attaining future objectives. In a time-horizon of 50 years, it is possible to relax many of the constraints imposed by today's infrastructure and practices. We can, for example, adjust our perspectives on groundwater and rain. We can be more efficient and responsible in our daily use of water, restricting waste and avoiding unnecessary pollution. Sources of contamination in the water system itself can be under control by 2040. Only organic materials will enter the system so that sewage sludge can be employed as a useful raw material.

A group of sustainability experts was brought together to sketch a vision of how a sustainable water system might look. The experts were helped in their task by the results of the strategic orientation analysis that had pinpointed key criteria for environmental sustainability in the water system and shown where opportunities for moving to a more sustainable system might be found using a more differentiated approach to water management. The important considerations are listed in Box 7.2. Against this backdrop, a sketch of a sustainable future water system was drawn in the form of a 'future history' (Box 7.3). This was then used in a backcasting exercise to reveal problem areas and bottlenecks that must be addressed in the near term in order to reconcile the actual situation with the sketch of the future we want to achieve. These problem areas will form the basis for decisions on needed innovations and transformations, acting as catalysts to accelerate the process of sustainable development.

BACKCASTING

Four scientists used the sketch of the future as a guideline to present their ideas on current trends that need to be broken and on new patterns and approaches that should be developed in order to reach a more sustainable water system. They presented their ideas at a workshop to an audience of actors and stakeholders representing the various branches of the water industry and the water-using community. Their presentations covered four distinct themes:

▷   Location-specific management of groundwater

▷   Household use of water

▷   Waste-water and precipitation collection

▷   Waste-water treatment

LOCATION-SPECIFIC MANAGEMENT OF GROUNDWATER

The essence of the approach is to make a sophisticated management of the naturally provided underground infrastructure based on a deeper understanding of local hydraulic regimes. The opportunity arises because the substratum is made up of several distinct hydrological systems. In effect, groundwater is separated naturally into relatively isolated reservoirs. This makes it possible to think in terms

WATER MANAGEMENT AND SUSTAINABILITY

▷ Water is both drawn from and returned to the environment. The more that is drawn, the greater is the desiccation threat and the higher is the energy and resource requirement of running the system. More efficient use of water, which reduces the need to draw on stocks in the first place, will improve sustainability.

▷ At both ends of the 'chain' water must be cleaned. Treatment following contamination imposes environmental costs, which increase with the quantity of contamination and its dilution. Treatment is never 100% effective. For sustainability, it is better to avoid dilution whenever possible.

▷ By contaminating the organic waste matter with heavy metals, the sewage sludge produced at water treatment facilities becomes contaminated and unusable. For sustainability it is essential to avoid heavy metal contamination.

▷ Different uses for water do not necessarily require water of the same quality. There is an issue of 'fitness for use'. For sustainability it would be better to develop a usage hierarchy.

▷ Centralisation implies that potable water and waste-water are transported long distances. For sustainability, a decentralised arrangement would lower energy use.

TODAY'S ARRANGEMENTS

▷ Today's water system is a general system. It makes no differentiation or separation between water of different qualities beyond that between potable and waste-water.

▷ Today's water treatment system is expensive, yet it is only 50%–60% effective.

▷ Two-thirds of the critical heavy metal contamination in today's waste-water comes from sources within the water system.

▷ Rainwater constitutes one-third of the total volume of waste-water.

▷ There are important interdependencies and interactions between the water industry and water users. Operational control is fragmented across the system.

OPPORTUNITIES OFFERED BY DIFFERENTIATION

▷ Underground water in the substratum is naturally separated into different reservoirs that could be used for separate purposes and separately managed.

▷ Context-specific solutions (tailor-made to the local situation) would allow local needs for potable water and waste-water treatment to be met locally.

▷ Rainwater falling on roofs could be collected directly and used for low-grade applications.

▷ Water could be cascaded down a usage hierarchy to reduce water demand and ensure a better fit between the qualities of water needed and supplied for each function.

▷ Different grades of waste-water could be collected separately and treated according to the contamination level.

▷ By keeping heavy metals out of the system, sewage sludge could be used productively.

Box 7.2 **Considerations in sketching a future vision for the water system**

66 A municipal water chain has many more aspects and is more differentiated today than 50 years ago. The uniform central system with one quality of usable water, one waste-water system and one big sewage treatment system to which houses, industries and roads were connected has developed into a geographically and functionally differentiated system in which water supply and treatment are better connected to quantity and quality aspects of need and to spatial aspects such as where the demand for water or waste treatment arises.

The water chain in the year 2040 is characterised on the supply side by a cascade of functions, with the quality of water finely tuned to the need and function. Waste-water from one function that is suitable for another (lower) function is cascaded down the function hierarchy. The water system is no longer used as a disposal and transport system for waste materials that are not intrinsic to the water system. Wherever possible, waste materials are held in solid form and are not put into the water system at all. When it is still necessary for wastes to enter the water system, as little water as possible is used to carry the waste away since this minimises the subsequent task of separation and treatment of the water and the solid wastes so that these can be re-used.

These changes have led to a separation of the different streams of water, dependent on the type of use. Systems hardware and installations for storage, transport and treatment are now being optimised to the density and type of use and according to space availability both within buildings and in different geographical areas. More space is allocated now for storm-water storage. In buildings, storm-water is used for low-quality applications, such as laundry, where soft water (without magnesium and calcium) has advantages. Towns now have different forms of storage and buffering capacity to deal with excess storm-water and run-off. In the suburbs, space is allocated to facilities that combine both storm-water storage and purification functions and these operate on a continuous basis. This lowers overall treatment costs since storage is less costly than adding peak load treatment capacity. The latest treatment systems are energy-efficient and reach a high performance. 99

Box 7.3 **The water system of 2040: extract from a 'future history'**

of managing each reservoir independently in respect to functions allocated to it on the basis of the quality of the water, the degree of isolation of the reservoir and the situation at the surface. Management of underground water would seek to improve the compatibility between conditions at the surface (land use type, local water needs, etc.) and the functions allocated to the groundwater. In urban areas, for example, groundwater could often serve as the basis for a 'second-quality' water network for functions such as washing clothes, flushing toilets or watering gardens. The overlying soil can also be used to filter storm-water and some household waste-water. Up to a certain level, impurities and wastes can be removed from storm-water and waste-water by filtration through soils and substrata and can even be broken down slowly within the groundwater system.[5]

5   The general ideas explored here also hold application possibilities beyond the domain of water management *per se*, some of which were explored in other STD projects: for example, the use of groundwater for heat storage was explored within the nutrition sub-programme.

HOUSEHOLD USE OF WATER

The essence of the idea is to differentiate the functions served by water in households and to cascade water down a hierarchy of uses. The current daily per capita level of potable water use in households is almost 160 litres. About 40 litres of this (25%) is used for bathing, showering and the washing of hands and faces. Similar amounts are used for washing clothes and for flushing toilets. If these three functions were coupled, so that water is cascaded down a usage hierarchy, there is scope to reduce the demand for potable water by 50%. This would also halve the amount of waste-water from households. Coupled with the elimination of storm-water from the sewer system (which currently constitutes about one-third of the total waste-water stream), this would reduce the overall volume of waste-water by two-thirds, increasing the organic waste concentration of waste-water received at treatment facilities by a factor of 3. In turn, this is necessary for enabling sewage to be treated by anaerobic digestion, which produces lower volumes of sewage sludge than aerobic treatment.

WASTE-WATER AND RAINWATER COLLECTION

The different grades of waste-water and sewage should be collected separately. A distinction should be drawn between precipitation, lightly contaminated waste-water and heavily contaminated waste-water. By separating the streams, different treatment regimes can be applied that are appropriate to the nature and severity of contamination. Heavily contaminated waste-water, such as from dentists' surgeries and photographic shops, should be held separately, collected and sub-jected to specialised treatment at centralised facilities. In contrast, most household waste-water and storm-water could be treated locally, saving on infrastructure and operating costs. Eliminating storm-water from sewers would also avoid the overloading problem at sewage works. Storm-water that runs off suburban streets can be returned to the ecosystem without treatment, whereas that which runs off heavily used roads and motorways must be collected and given a simple physical treatment to remove contaminants. Household waste-water needs a biological treatment, but this could be performed locally using anaerobic processes if the concentration of organic matter in the waste-stream is three times higher than today. Chemical and physical treatments may also be needed to precipitate heavy metals.

WASTE-WATER TREATMENT

The technological needs and opportunities in respect to small-scale facilities for water treatment within the built-up areas of cities, towns and suburbs need to be explored through R&D. Physical, biological and chemical treatments are in principle available to be improved. Especially, there is a need for R&D on anaerobic treatment processes for household waste-water as this would reduce the amount of solid sewage sludge produced and, in principle, could also yield methane as a useful by-product. Anaerobic treatment of sewage converts the organic waste content of the waste-water to a mixture of solids and gases. The sewage sludge produced during anaerobic treatment is equivalent to only 5% of the original mass of organic waste. The rest of the organic matter is converted to gases. Anaerobic bacteria can co-exist with some heavy metals and will accumulate these and other toxic materials, such as pesticide residues. Anaerobic treatment therefore offers a way of treating waste-water that contains metal and pesticide contaminants. These cannot be completely broken down by the bacteria so the sewage sludge is con-taminated. However, relatively little solid sewage sludge is produced, which opens up options for final disposal that are not possible with aerobic treatment, such as vitrification of the sludge. The cleaned water-stream will be of a higher standard than if aerobic treatment processes are used. Before being returned to rivers, the cleaned water-stream can undergo a precipitation step to remove remaining metal contamination and a freshening step to raise the oxygen content.

Anaerobic treatment is already used today for some industrial waste-water streams, especially in the potato starch and sugar-refining industries. However, under today's technologies, anaerobic treatment requires a high concentration of organic matter in the waste-water and a temperature of over 30°C. The waste-water stream entering sewage works today has a low concentration of organic matter and a temperature of around 15°C. Currently, then, only aerobic treatment processes are used at sewage plants. In order to use anaerobic treatments, several organisational and technological breakthroughs are needed. The first is to increase the organic concentration of the waste-water stream: for example, by water cascading and by eliminating storm-water from the sewerage system. Another is to develop anaer-obic bacteria that function well at low temperatures. A remaining technological issue is how to separate the solid, liquid and gaseous components produced as outputs from the anaerobic digester. Gases leaving the digester will contain some solids that should be returned to the digester for further treatment. A promising approach to this technical problem uses a funnel system to separate gases, liquids and solids, using the differential velocity and viscosity of the various materials to make a one-step separation. This area needs further R&D.

In principle, then, two different approaches to waste-water treatment are pos-sible. One approach, based on aerobic digestion, is suitable if heavy metal contami-nation can be kept out of the waste-water stream. This depends on switching from metal to PVC and PE pipes within the water system and preventing heavy metal contaminants entering the sewerage system from external sources. The other approach, based on anaerobic digestion, is suitable for treating waste-water that contains heavy metal and pesticide contaminants. This approach depends on increasing the organic content of waste-water and achieving compatibility between the temperature of the waste-water and the temperature needs of anaerobic bacteria. The two approaches should be developed in parallel, as both are likely to be needed in the future in different application contexts.

USE OF LCA IN THE DEFINITION PHASE

On the basis of the ideas presented by the scientists and critiqued during workshop sessions, five measures that might contribute to a sustainable water system were defined and a life-cycle analysis performed to calculate the results that could be achieved using each measure. The LCA took into account all of the previously identified environmental effects. The following measures were formulated:

▷ Cascade systems are introduced into households to reduce the use of potable water.

▷ Precipitation is no longer collected in the combined sewer system but, wherever possible, is infiltrated into the soil via a system of ponds.

▷ All water pipes, roof gutters and drainpipes made out of zinc, copper or lead are replaced with PVC or PE.

▷ Precipitation that falls on traffic arteries is collected separately and treated separately rather than being allowed to run off into the combined sewers.

▷ Treatment of waste-water from suburban houses and residential areas takes place locally in small-scale treatment plants.

In order to evaluate these options, it was necessary to contextualise them using an appropriate case-study region. A region typical of Dutch conditions was chosen in the east of the Netherlands: the region surrounding Almelo and Wierden. The life-cycle analysis revealed that, while each of the five options would indeed bring about significant improvements, no single alternative in isolation would accomplish a reduction of the environmental effects by a factor of 20 (Figs. 7.3–7.7). A combination of innovations was likely to produce a more significant result with much bigger jumps toward sustainability. The various measures were, therefore, analysed in different combinations. In order to make this evaluation, the case-study region was broken into five separate parts so that sets of measures in combinations appropriate to the specific applications context of each part could be modelled. Such a tailor-made solution based on sets of integrated measures was found to achieve a far greater level of environmental improvement. Moreover, because of interaction effects between measures, synergistic improvements in environmental performance can be obtained.

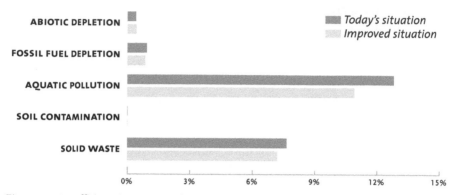

Figure 7.3 **Eco-efficiency improvement by water cascading**

Source: TNO-MEP 1997

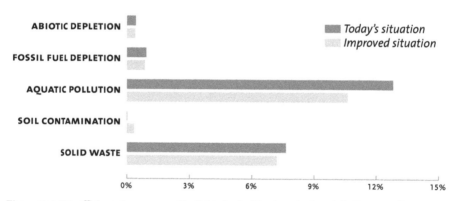

Figure 7.4 **Eco-efficiency improvement by light physical treatment of precipitation run-off**

Source: TNO-MEP 1997

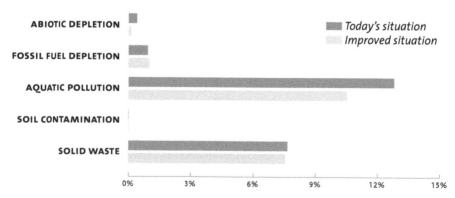

Figure 7.5 **Eco-efficiency improvement by excluding heavy metals from infrastructure**

Source: TNO-MEP 1997

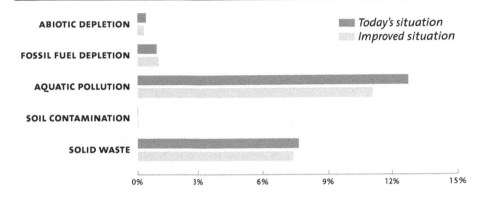

Figure 7.6 **Eco-efficiency improvement by separately treating contaminated precipitation**

Source: TNO-MEP 1997

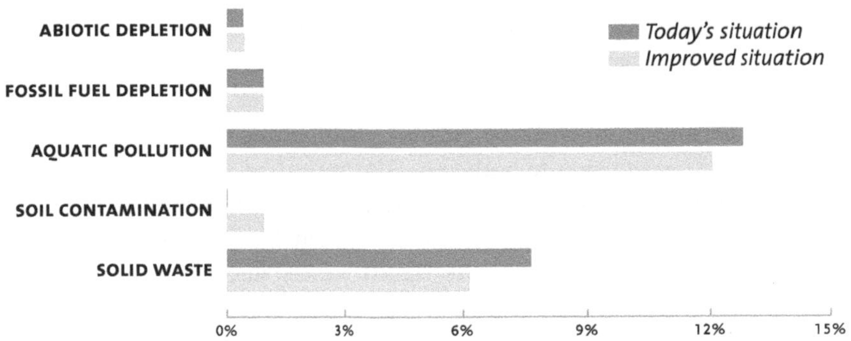

Figure 7.7 **Eco-efficiency improvement by locally treating household waste-water**

Source: TNO-MEP 1997

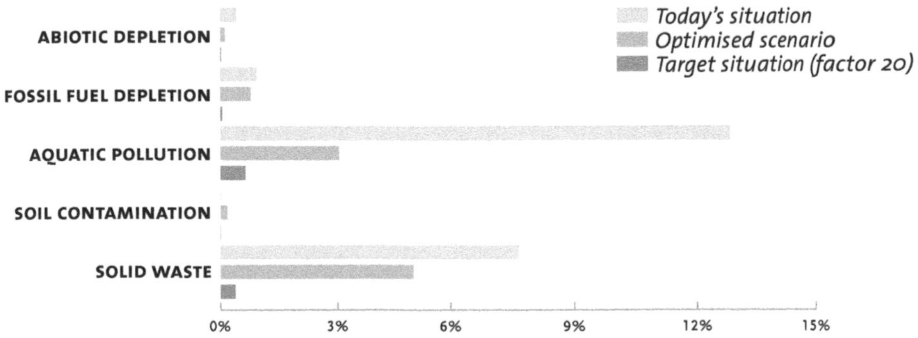

Figure 7.8 **Eco-efficiency improvement achieved by an integrated solution in the Almelo case study**

Source: TNO-MEP 1997

In essence, synergistic effects arise because the effectiveness of each measure is conditional on other measures being implemented. The efficiency of treatment installations can be greatly improved—for example, by pollution prevention at source—but the benefits at treatment plants can be maximised only if the dilution factor is also reduced and if inert materials are used in pipes and gutters. Equally, the benefits from separating waste-water streams are only fully available when cascades are implemented in households, which makes it possible to treat waste-water in local facilities. An interesting conclusion is that the most of these measures are implementable today. They depend less on technological innovation than on organisational innovation. Figure 7.8 presents the normalised environmental scores for a restructured water system. It represents what could be achieved in the test region of Almelo by optimally combining an integrated set of measures. An integrated solution could reduce abiotic depletion by a factor of 3, fossil energy use by 15%–20%, aquatic ecotoxicity by a factor of 4 and solid waste production by 25%. There is an added benefit in that the heavy metal content of the solid waste is also reduced. These gains are partly offset by a small increase in direct contamination of soils associated with no longer treating all storm-water.

Figure 7.8 also compares the environmental performance achievable through this optimised arrangement with the current situation and the factor-20 target of the STD programme's working hypothesis.[6] The optimised arrangement achieves a considerable improvement over today's system, but still falls short of the factor-20 target. The main reason is that the causes of remaining problems cannot be tackled by measures taken by the water industry itself since they are created by practices outside of water industry control. In effect, while measures available to be introduced today can be applied and will reduce the environmental strain, a sustainable water system can only become a reality if changes are also made in other sectors. The net needs to be cast much further in order to tackle contamination entering the sewerage system from traffic, agriculture, industry and households. Some of these external sources are being addressed in other parts of the STD programme and the links with STD projects in such areas as nutrition, housing, transport and chemistry form a useful starting point for a cross-sectoral approach. The use of sustainable energy and renewable raw materials will contribute to a further reduction in environmental strain as, also, will the development of more sustainable agricultural practices.

IMPLEMENTATION

The idea of developing a support base is central to the entire STD programme. In the water project it was clear early on that a local implementation network would

6   Quantitative data on desiccation was not available and therefore cannot be included in this comparison.

add value to any results and that a communications process was needed in parallel with the analytical activities. A 'sounding board' was established in the course of the Almelo case study made up of representatives from the municipality and the various agencies involved in the water chain. By means of intensive interaction and communication, the level of interest and involvement of members of the sounding board was sufficient for the activity to be carried forward autonomously. At the point when the STD programme stopped its activities, members of this local network took up the ideas from the project. An agreement has been reached between the region's water authority, the district councils, the local authorities and the water companies to make a concerted effort to create a sustainable municipal water system between now and 2040 and to implement a strategic plan for the sustainable management of ground and surface water. The agreement is in the form of a binding 'pact' made between the various parties and signed by them.

The agreement overcomes the critical obstacle to strategic action—that individual actors lack assurance that stand-alone measures will yield benefits sufficient to justify the expense involved. This is especially important when benefits of individual measures taken by one party depend on measures taken by others, especially if the various measures are implemented in different time-horizons. In Almelo, some measures need to be phased in over several years and will only be implemented when replacement investment is needed in any case. An example is the separation of rainwater run-off from roads. Other measures can be implemented immediately and can be stimulated through subsidy programmes and grants. An example is the elimination of zinc, copper and lead in pipes and gutters of private houses.[7] By making the necessary actions part of an agreed plan, the decision calculus of individual actors has thus been changed in favour of making strategic choices.

CONCLUSIONS

In what ways has the STD water project been useful? And what role did LCA play in achieving a positive outcome? The major contribution of the water study is that it clarifies that unsustainability is intrinsic to today's approach to managing water resources and liquid wastes. Today's model, based on a general system that recognises only two qualities of water, represents an end-of-pipe approach. This necessarily introduces an incompatibility between different sustainability criteria,

---

7   Water cascading within homes has been calculated by the project to be just feasible economically. The main costs are in storage space, filters and energy. The main benefits are in reduced use of potable water for low-quality needs such as flushing toilets. The approach depends on installing a secondary storage system and reorganising the use of space within buildings. Again, grants and preferential loans are prospective policy instruments.

ensuring that improvement on one is automatically offset by deterioration on another. Even pursued to its ultimate extreme, today's approach could never offer a route to a sustainable water system. The STD study provided opportunity to make the necessary reappraisal of the entire water system on the basis of combining sets of measures that, individually, have little impact, but work well in combination.

Within the project, LCA was used not only as an instrument for evaluating alternative water system designs, but also as an instrument for project steering. The foundation of the approach was a decision scheme. Each LCA was scrutinised to see whether there had been a jump in environmental performance. This was done at several different levels: for individual measures, subsystems, linked sub-systems and fully integrated solutions. Where there was no jump, that development trajectory was abandoned. Where there was a jump, it was asked why. LCA was also used to indicate and explore practical problems that would be encountered in the implementation phase. One key problem is the number of different parties whose actions need to be co-ordinated. Another relates to the balance of costs and benefits of various measures taken in isolation or in combination and to the distribution of these across actors and in time. To facilitate a strategic perspective on the future development of the water system, an outline agreement was designed by which the various parties could promise each other that they would implement different parts of an agreed plan as these fall within their domain of responsibility. This made it possible for the findings of the project to be used in the case-study region around Almelo as the basis for a strategic development plan. Since the Almelo Pact was signed, several other regions have taken up the results of the project and are now implementing their own plans using the template of the Almelo case study.

Chapter 8
Gaining 'virtual hindsight' by backcasting: the case of service products

This chapter is based on work by Philip Vergragt, Marjan van der Wel and Robert van den Hoed (Vergragt *et al.* 1995; Vergragt and van der Wel 1998; van den Hoed and Vergragt 1996). It concerns the links between cultural, behavioural and technological innovation and between the different time-horizons for innovation. The chapter explores two parallel opportunities for reducing environmental burdens. One lies in need reduction. To what extent have today's expectations—as reflected in demands for services—grown through co-evolution with facilitating technologies? Will it be necessary to meet all of today's demand in the future or can we envisage strategies for demand reduction? The other opportunity lies in changing the ways in which remaining needs are met. Rather than the consumer buying, owning and using small-scale service-providing machines, we can instead envisage a more eco-efficient solution that involves the consumer buying services directly from a dedicated service provider. The particular case examined by the STD programme concerns the 'need' for clean domestic textiles.[1] Is there scope to reduce the environmental burden of textile cleaning by outplacing the laundry function to professionals and upscaling the function? Can we replace household-scale washing machines and tumble-dryers, which place heavy demands for electricity, water and detergents, with eco-efficient laundry services? This introduces the notion of a service as a product in its own right. While this chapter concentrates on textile

---

1   Initially, the STD laundry project was started as a joint initiative involving representatives of different departments of the Technical University of Delft in association with the STD programme. An interdisciplinary group was formed with representatives from gender studies, engineering, technology management and several other academic disciplines within the Technical University.

cleaning as a specific case study, its wider implications concern the scope for upscaling several other functions now performed eco-inefficiently within households.

The service product and demand reduction strategies provide important opportunities for reducing the environmental burdens imposed by households. However, successful implementation requires a set of interdependent, co-evolutionary innovations in culture and behaviour as well as in technology. In order to initiate a new innovation trajectory, we need to be clear about where we are heading, what actions are needed, when these need to be taken and how actions are interdependent. Forward planning in the short and middle terms calls for 'virtual hindsight'. One way to gain this virtual hindsight is to develop a normative image of a desirable future state and to backcast from this, writing an imaginary history of how an acceptable state of affairs was reached. The history should encapsulate the interdependencies between innovations of different kinds and show how bottlenecks were successfully tackled or avoided. Using one or more of these histories, we can draw implications for initiatives that need to be started now to have any chance of making the future happen as we have envisaged.

This approach, which was used in the laundry case study, makes use of a specific variant of backcasting methodology called the 'future perfect'. The aims of this chapter are to describe the use of this backcasting method and the results obtained by using it. The main finding from this project is that there are potentially interesting laundering technologies that could be used in the future to clean household textiles and that, by using them, there is scope to improve eco-efficiency, but also that these technologies are not appropriate for use on a household scale and their eco-efficiency benefits are, in part, a function of upscaling. In particular, in the case of continuing with water-based textile-cleaning technologies, upscaling offers scope for optimising the washing processes and for achieving a precise match between washing needs and washing routines. Water, heat and cleansing agents could be recovered and re-used. In the case of using alternative, non-water-based cleansing technologies, the nature of the technology and the level of operator expertise required mean that these could be offered only on a professional scale. The long-term development of these improved technologies depends on establishing a level of demand for professional laundry services in the short to medium terms that is necessary to drive the R&D process. The future perfect method provides useful insights into the bottlenecks involved and how these might be overcome.

STRATEGIC PROBLEM DEFINITION

Today's approaches to meeting needs are the outcomes of a co-evolution between technology, markets, culture and behaviour. Perceptions of 'needs' as well as the approaches taken to fulfilling these are a function of co-evolutionary processes, in which self-reinforcing and -perpetuating feedbacks often dominate. Perception of need influences the technological possibilities to satisfy need and is influenced by those possibilities. Thus, for example, having affordable means of personalised transport has influenced the perceived need for mobility. Cars make travel easy and so encourage patterns of land use and lifestyles that generate demands for mobility. The outcome is that the technology reinforces the original need as well as meeting it. This is one of several so-called 'rebound' mechanisms by which technological innovation on the supply side can lead to an expansion of needs and wants on the demand side, so undermining some of the potential societal benefits of efficiency gains and cost reductions.

The need for textile cleaning is a case in point. Concepts of hygiene and cleanliness as well as approaches to cleaning domestic textiles are outcomes of the technologies developed for performing the laundry function. The perceptions of need, the product and process technologies used, and behaviours in respect to meeting 'laundry service needs' are co-related through their co-evolution. The fundamental assumptions, principles and design criteria that underlie the technologies thus come to assume an exaggerated significance, because the technology, together with all its advantages and disadvantages, becomes part of a complex and interdependent 'system'. The system represents an approach to fulfilling the need that is more than simply technologically and economically defined. Cultural and behavioural aspects also define the system, including both the means and the ends. Patterns of behaviour and perceptions of need co-evolve with the technology. If the technology is effective, convenient and affordable, it is likely to become widely used. If, at the same time, the technology has a high environmental impact per unit of delivered service, its widespread use may begin to impose a high absolute demand for limited eco-capacity.

The problem will be more difficult to deal with because of the interdependencies. Over time, the system as a whole and not just the technology becomes habitualised. To reduce the environmental burden, the whole system will need to be restructured, not just the technology. Against this backdrop, it is important to revisit the fundamental assumptions and principles on which today's technologies and systems for meeting service needs have been developed and to rethink these from alternative perspectives. This is the case with laundry services. In order to initiate a new innovation trajectory, the limits of the prevailing trajectory and the scope implied by alternatives need to be explored together with the stakeholders and actors who have the power to envision and create an alternative future.

TODAY'S APPROACH TO TEXTILE CLEANING

Before discussing new solution approaches, it is important to clarify the premises that underlie the dominant approach to textile cleaning. The starting points that characterise today's approach and that have been steering its development through-out the past decades are:

▷ The use of water and detergents as textile cleansing agents

▷ The positioning of the laundry function within the informal household sector of the economy with only a few, specialised laundry tasks being transferred to the formal sector

▷ The design and optimisation of the washing machine at the scale of the household

From inception to the present day, there has been no fundamental change in the basic design of the household washing machine, albeit that the design has been incrementally improved and optimised. The domestic washing machine today is still a tub-drum apparatus in which dirty textiles are agitated together with water and detergents.

Possible new solutions to serve the need for textile cleaning can be based on a reassessment of these basic starting points. Figure 8.1 maps out a solution field in two dimensions using two of the three criteria implied by the starting points just listed. The two dimensions are the extent to which the activities are profes-sionalised or 'outplaced' to the formal economy (vertical axis) and the scale of the technology (horizontal axis). The scale of the technology can be translated into various terms: the size of the technical device, the number of people using the same technical device or the average distance of the technical device from those who use

Figure 8.1 **The position of the existing household washing machine in the solution field**

it.[2] A third dimension (not represented here since it is not continuously variable) is the nature of the technology used for removing dirt and stains. To some extent, the nature of the cleansing technology is not independent from these other two dimensions. Some chemical cleansing technologies can only be used safely and properly by experts, which implies outplacement to a professional. In turn, this is likely to imply a larger than household scale for the technology. Professional dry-cleaning services that use organic solvents and operate in local neighbourhoods are cases in point today.

The existing technology of the household washing machine can be placed in the bottom-left side of the solution field: we mostly do the washing ourselves and we use a technical device that is owned by and located within the household in which we live. Other arrangements that exist to varying extents today (although none is particularly well represented) include the employment of professional house-keepers within the household (top-left side of the solution field), the use of com-munal appliances within apartment buildings and at launderettes (bottom centre), and the use of laundry and dry-cleaning services (top right). By exploring the remaining areas of the potential solution field, we can describe other plausible arrangements and elaborate on the scope for improving arrangements that already exist but are rather under-represented today. Even when future images are a reinterpretation of known, old practices, they can take on new qualities if needs are fulfilled using new technologies.

ENVIRONMENTAL ASPECTS OF TODAY'S APPROACH

The cleaning of textiles is a source of environmental pollution today because of its draw on electricity, water, detergents and related materials such as water softeners, anti-foam agents and surfactants. These resource implications of operating the laundry equipment are more important from an environmental perspective than the resource implications of having five million washing machines in Dutch society or replenishing this stock of physical equipment with upwards of 500,000 new machines annually. This is because up to 90% of the materials and components embodied within washing machines is already re-used or recycled. The recovery process has an energy cost. However, this represents only a small percentage (less than 8%) of the total annual energy use associated with performing the laundry function. Including the recycling energy cost in the analysis, the energy use associated with today's domestic washing machine technology is 0.45 kWh per kilogram of washing. The comparable figure for water is 25 litres per kilogram and for detergent use is 45 grams per kilogram.

---

2   The issue here is whether we can reap 'eco' (eco)nomies of scale. That is, whether we can increase resource productivity by increasing the scale of the service-delivering process and machinery and the intensity with which these are used.

Returning to the wider concern of the sustainability of households and the possibilities for upscaling as a way of improving the situation, it is important to contextualise washing within the wider trends that have been affecting households over the past decades. The present household is becoming more and more like a small factory with virtually every household task now automated. The development and implementation of the household-scale automatic washing machine and tumble-dryer provided an answer to the time-consuming task of manual washing. At the same time, the frequency of washing has strongly increased while the notion of what passes for hygienic has tightened up. In effect, washing now seems not only to fulfil a material need, but perhaps a ritualistic and symbolic one as well. The average annual demand for clean laundry in the Netherlands is now around 230 kg per person, which is equal to 690 kg per household. Textile cleaning now accounts for much of the environmental stress generated by households. It accounts for around 20% of total household water use and a similar (but growing) percentage of total household energy use. The absolute total environmental stress caused by households is likely to increase if the present trend toward using domestic labour-saving devices continues.

THE FOCUS OF CURRENT R&D ON TEXTILE CLEANING

Incremental R&D in respect to the current approach to textile cleaning is focused on reducing the energy, water and detergent use of household washing machines and dryers. However, the scope for further development along these lines is limited. A recent study of the potential for environmental improvement of the existing household washing machine technology conducted by the Group for Efficient Appliances identified several design improvements that are, in principle, feasible and would reduce the environmental burden. These include improved water level control, improved thermal efficiency, reduction of tub-drum clearance, extended motor efficiency improvement and automatic dosing systems. Table 8.1 gives findings about the scale of the eco-efficiency improvements likely to be reachable by such improvements. The analysis is performed for energy and water taking account of both best available technology (the standard base-case) and average in-service technology (real-life base-case). The study compares today's efficiencies with the efficiencies that might be reached using improved technologies. Improvements are classified according to whether they are technically feasible and of economic interest for machine manufacturers to implement, technically feasible but not of economic interest to machine manufacturers and experimental. Only technically feasible and economically attractive improvements are likely to be realised through autonomous technology development. The study concludes that there are strict limitations to the eco-efficiency improvement potentials of autonomous development of household washing machines, especially if these are compared against the factor 10–20 goal of the STD programme.

	STANDARD BASE-CASE	REAL-LIFE BASE-CASE	STANDARD BASE-CASE	REAL-LIFE BASE-CASE
	Energy (kWh/kg load)		Water (litres/kg load)	
Current situation	0.30 100%	0.45 100%	18.5 100%	24.5 100%
Economic	0.23 77%	0.21 46%	14.6 79%	12.0 49%
Technically feasible	0.20 67%	0.17 38%	11.4 62%	11.0 45%
Experimental	0.17 57%	0.15 33%	10.6 57%	11.0 45%

Table 8.1 **Long-term efficiency targets for domestic washing machines**

Source: GEA 1995

With regard to clean textiles, as with regard to many other areas of need, incremental technological improvement based on technical adaptations of the existing ways in which needs are fulfilled gives insufficient scope to improve eco-efficiency. Today's research efforts are focused on improving conventional domestic appliances rather than on developing new concepts of how to fulfil needs. In effect, this is to treat the problem as a purely technical one as if it were amenable to a purely technological solution. In practice, a technological solution is insufficient. A sustainable solution will depend on systems-level changes that involve changing the way that technology is used as well as changing the technologies themselves.

Technological development within the 'service product' approach is the result of taking a new definition of the task at hand based on 'need fulfilment' as the starting point for R&D. In the case of clean textiles, the starting point is that people do not need washing machines, but only clean clothes and clean household linens. Although this may seem so self-evident as to be trivial, this is not the case. Taking the need for clean clothes as a starting point for R&D is fundamentally different from assuming that people need household washing machines. The different starting points lead to fundamentally different R&D agendas and to the use of fundamentally different sets of technology assessment criteria. The way in which the need is fulfilled and the values that are achieved by fulfilling the need are also different between the approaches. In the current situation, the transaction value of the appliance at the point of sale is the most important value. The manufacturer focuses on this and seeks to optimise the product on price and technical qualities in respect to the one-time purchase decision of the eventual owner and user. At the point of sale, save for a short period of warranty, all costs and risks arising from the performance, operation and maintenance of the machine during use as well as from disposal at the end of its useful life pass to the consumer and to society as a whole. By taking the need that is to be fulfilled as a starting point, the focus of R&D can be shifted to entirely new solution directions and to new qualitative aspects of the way that the service is provided.

The demand for textile cleaning is a function of the quantity and characteristics of the textiles in use, the ways that textiles are used (or not used), the design of textile products and people's perceptions of cleanliness and hygiene. These imply that several potential strategies are available for reducing the environmental burden of textile cleaning. Among others, these include demand reduction, textile redesign and upscaling of the laundry process.

▷ In principle, demand reduction could be achieved by shifts in people's perceptions of how often clothes and other textiles need to be cleaned or by redesigning textiles to be fundamentally more dirt- and stain-resistant. There is also the possibility of designing clothing such that those parts that become stained can be detached and cleaned without necessarily cleaning the whole garment.

▷ Substitutes for textiles could also be considered such as using blinds and shutters in place of curtains or wearing disposable rather than conventional garments. Fibres and textiles could be redesigned so that cleaning and finishing (drying and pressing) place lower demands on eco-capacity than they do at present. By using fibres that absorb less water during washing, for example, less energy is needed for drying.

▷ Upscaling could, in principle, increase the environmental efficiency by enabling both conventional cleaning processes to be optimised as well as fundamentally different cleaning technologies to be used. Upscaling might allow recovery and re-use of water, energy and cleaning agents or permit precision control of washing processes in relation to need. It could also allow the use of cleaning processes that are not based on using water and detergents but on supercritical solvents or biological cleansing agents.

At first glance, the evaluation of these alternatives may seem straightforward. However, making the transition from doing laundry at home to sending it out for professional laundering is not so simple. Shifts in culture, behaviour and technology are implicit in upscaling. There are also logistical and organisational implications because of the need to transport textiles between their owners and centralised laundry facilities and to sort textiles according to cleaning requirements. These shifts raise many questions. What are the technological possibilities and the time-paths for their development and introduction? Will a shift to buying laundry service rather than doing laundry at home appeal to consumers? How much are consumers willing to pay? What other aspects of laundry service provision will be relevant to consumer choice? Will a shift to the provision of laundry services reduce environmental impacts significantly? How big a shift in behaviour is needed and by when to realise significant environmental gains? What sorts of policy are required?

THE STUDY APPROACH
AND ORGANISATION

The approach taken by the project to answering these questions was a variant of the general STD methodology that put special emphasis on interactive Constructive Technology Assessment and on the use of the 'future perfect' backcasting method. The interdependencies between cultural, behavioural and technological innovations implied by a shift to service products means that a strong emphasis must be given to involving representatives from all the key stakeholder groups. Progress is made by alternating phases of discussion with phases of research and by working iteratively so that decisions are constantly reviewed and reconsidered in the light of new knowledge gained during the project. The future perfect methodology affords a special role to the future history as an instrument for achieving project objectives. The future history is used to forge a link between the inductive and deductive work of the project. In the beginning (a largely inductive research phase), efforts are focused on proposing and evaluating possible new approaches to meeting needs. Results are summarised in the form of a sketch of a desirable future situation, which represents a coherent 'vision' of a plausible and desirable future. This is then translated into a set of future histories, which show how this future situation was reached. The future histories in turn are used (in a largely deductive way) to draw inferences about short- and medium-term actions that will be needed to support and inform the implementation process.

The study involved a project management team acting as facilitator and a group of stakeholders and actors. From the outset, the project was conceived as necessarily involving stakeholders from at least three different groups: producers, consumers and technical specialists. Industrial and business participation came from several important equipment and detergent manufacturers, dry-cleaning companies, textile producers, textile leasers and laundry operators.[3] The participation of stakeholders from business and industry was especially critical because

3   Obtaining industrial participation is not easy, especially where confidentiality considerations play an important role. Industrial participants were approached and encouraged to join the project through a specialist industrial consultant with interests in environmental and consumer aspects of technologies. The consultant acted as an intermediary in respect to a large network of industrial stakeholders. From the perspective of all involved, this represented a win–win situation. The consultant saw this as an opportunity to serve his client base and industrial stakeholders were more inclined to join in the project knowing that it had been recommended to them by a trusted figure.

of their specialised technical knowledge and their potential role in reorienting business strategies and R&D programmes. Representatives of the different consumer groups and housewives' organisations took part directly. In addition, the research work involved making a questionnaire survey of consumers. Technical specialists from several research facilities and universities were also involved. An important task was to make a social map of stakeholders, including the potential interests and aims of each in the project and where these might coincide or conflict. As part of the process, stakeholders were asked explicitly about their interests in the project and expectations of it. A selection of answers is presented in Table 8.2.

Most participants expected to gain from the exercise as well as to contribute to it. For many, the opportunities to develop insights into new innovation trajectories and build new contact networks were clearly important. Several of the industrial and commercial participants saw potential business opportunities in upscaling and outplacement of the laundry function. But by no means everyone was convinced that this offers a promising way forward. Several participants were rather sceptical of the environmental benefits from upscaling and wanted to see whether, indeed, the project would show whether such benefits are available.

Two main workshops were held, separated by one year. Prior to the first workshop, preliminary work was undertaken to establish the current situation and the likely consequences of an autonomous development trajectory. The first workshop was essentially a creativity session at which the problem of cleaning textiles was revisited from first principles and new ideas were solicited from stakeholders about potential solution pathways. The period between the two workshops was an inventory and research phase during which the ideas sketched out in the first workshop were worked up into scenarios, fleshed out for reaction and built into a future history. At the second workshop, the future history developed during the first year, which represented the synthesis of findings about promising new innovation possibilities, was used to structure a targeted discussion aimed at developing project proposals, work plans and co-operation among innovation agents. The second workshop thus marked a pivotal point in the transition from background research to an implementation and installation phase.

BACKCASTING USING THE FUTURE PERFECT METHODOLOGY

In this section and the one that follows we consider the substantive outcomes of the two workshops and the future history that was developed and evaluated in the interval between them. The results include descriptions of several different approaches to textile cleaning, evaluations of these from environmental, economic and consumer acceptance perspectives and a set of research agendas covering the

PARTICIPANT	POSITION, ROLE OR RESPONSIBILITY	INTERESTS AND EXPECTATIONS
Major washing machine manufacturing company that exports to the Dutch market	Director of technical services responsible for sales and after-sales in respect to domestic and commercial appliances, e.g. washers; dryers and dishwashers for homes; laundries and hospitals	Interest in reducing the environmental burden; seeks clarification of whether gains can be made by bigger economies of scale and of implementation possibilities
Major washing machine manufacturer	Head of washing technique section that markets powders and develops systems for delivering powders during the wash cycle	Keen to formulate concrete research projects; to professionalise the sector by means of research and networking
Dry-cleaning company	Owner of the company and member of the board of a trade association of dry-cleaners	Interested in innovations for the dry-cleaning trade and in exploring the possibilities for association members
Textile-leasing firm with leasing and laundry operations in several European countries	Senior executive	Considers reducing the environmental load to be a key strategic challenge; interested in solution possibilities and implementation.
Manufacturer of ecological washing powder	Senior executive	Interested in future developments generally in respect to washing and the environment; to develop contact networks; to know about the environmental benefits of upscaling; to formalise concrete research questions
Direct marketing company supplying textile diapers	Owner	Interested in high-tech textiles with treated surfaces that are stain-resistant and easy to clean; aware of need to bring together professional and technical skills/knowledge from several areas to realise the possibilities for sustainable textiles/laundry; keen to contribute to this and to gain from it; sees large potential growth market
Organisation of housewives	Organisation member	Interested to ensure that, whatever solutions are proposed, these benefit households and housewives; to clarify what role housewives can play; to change mind-sets regarding hygiene
University of Amsterdam	Professional social-science researcher	Interested to maximise potential synergy of participants; to see how new possibilities are perceived by participants

Table 8.2 **Workshop participants and their interests in the laundry project**

technological and cultural innovations required for implementing the scenarios. The future history was based on a synthesis of the different approaches. An important aspect is that the different approaches and the scenarios developed in order to evaluate these should be considered less as alternatives and more as complements to one another. In effect, we are likely to see several different approaches becoming available to consumers, leading to differentiation in the ways that laundry needs are met in the future.

DEFINING NEW APPROACHES

As a backdrop to the first workshop, participants were presented with some preliminary ideas about upscaling, the solution field described in Figure 8.1, and results from a study of current laundry arrangements. Participants were also given an abstracting exercise to help them think creatively about sustainability issues. After these preliminaries, participants were divided into discussion groups and encouraged to use their knowledge and insights—old and new—as guidelines in considering the need for sustainable solutions to the textile-cleaning problem. Participants were then brought together again to discuss the findings of the different working groups. These could be structured around three key concepts. The first has to do with a better differentiation of textile-cleaning needs, the second with the possibilities for improving environmental efficiency through innovations of different sort, and the third with the complex connection between cultural, technological and organisational innovations.

At present, there is hardly any differentiation of textile-cleaning needs. Almost all needs are seen as identical and are treated identically. But needs could usefully be differentiated according to the types of textile and soiling. A useful distinction is between clothing and household textiles such as towels, bed linen and curtains. We can also differentiate large from small items and difficult- from easy-to-clean items. Another useful distinction is between sensitive and non-sensitive items defined on grounds of value, replaceability and intimacy. Useful distinctions can also be drawn between different types of soiling and whether all or only part of an item is soiled. This approach led to a suggested dichotomy in the organisation and technology of textile cleaning with a small, personal laundry at home for high-value, precious and intimate items and a specialised professional laundry service handling bulk textiles.

In order to appreciate the significance of this dichotomy, we need to look at how technological innovation could open up opportunities for improving the environmental efficiency of laundry processes. Based on the insights of workshop participants with specialised technical knowledge, several suggestions were made about product and process opportunities for improving environmental efficiency. These are reported in Table 8.3, which considers four steps in the conventional laundry process: pre-treatment, washing, drying and finishing. The opportunities for

		PROCESS IMPROVEMENT	PROCESS SUBSTITUTION	DEMAND REDUCTION
General				Use disposable textiles or textile substitutes. Use fewer clothes/textiles. Clean textiles less often. Make garments with detachable parts that can be separately washed. Treat local stains without washing entire garment. Develop non-staining and non-absorbing fibres and fabrics.
Pre-treatment		Develop stain-loosening techniques using biotechnologies		Develop fibre coatings that allow for easy stain removal.
Washing	*Water*	Wash using locally captured rain or 'grey' rather than potable water. Develop closed water systems.	Wash without water using organic solvent, supercritical CO_2, UV-light, bacteria, microwaves or other cleaning technology.	
	Detergent	Use bio-engineered enzymes. Target enzymes/detergents at specific stains using sensors to determine treatment/dosing. Recover detergents.	(as above)	
	Energy	Develop low-temperature washing processes. Use waste heat from other appliances.	(as above)	
Drying		Create built-in drying closets that dry using ambient solar/wind energy and/or waste heat from the refrigerator/ freezer.	No drying required	Develop fibre coatings so that fabrics hold less moisture and dry more quickly.
Finishing				Develop non-iron and crease-resistant fibres and fabrics.

Table 8.3 **Eco-efficient ways to meet the need for clean textiles**

improvement are structured in relation to these and according to how environmental stress is reduced: by improving the conventional laundry process, by substituting an entirely different process, or by demand reduction. In the future, these technologies could enable the laundry task to be split into two. A small home laundry for small and valuable items might be achieved by using local stain removers, a microwave, a thermos box or a simple process machine. The bulk laundry could then be outplaced for professional cleaning to a service using upscaled versions of water-based processes that allow for continuous process technologies or entirely new processes, such as ones based on supercritical CO_2 (Box 8.1). In addition, Table 8.3 envisages an important role for demand management in reducing the quantity of textiles that need to be cleaned and the frequency of cleaning.

THE SCENARIOS

After the first workshop, relevant criteria, parameters and ideas extracted from the discussions were structured into coherent scenarios, which were subsequently proposed to participants for comment and review. Figure 8.2 sets out three of the arrangements chosen for more detailed study and locates these within the solution field. In addition to today's concept of home washing, used as a benchmark, Figure

Figure 8.2 **The position of the scenarios in the solution field**

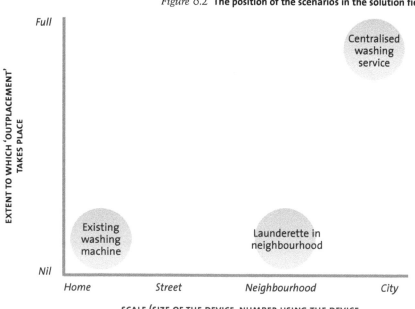

SENSOR TECHNOLOGY IS ALREADY BEING DEVELOPED IN SEVERAL APPLICATION areas. One area is textile cleaning. Sensors are already used to sense the weight of the wash, the water level and the water temperature within washing machines. These are simple applications. Sensors could additionally be developed to measure the cleaning needs of items, identify and locate stains of particular type, and feed information to dosing devices that would target appropriate cleansing agents on specific stains. Sensors could also be used to monitor washing processes, ending them when stains have been removed or changing washing conditions according to need.

Equally, enzymes and bio-engineered bio-catalysts are already used to increase the effectiveness and efficiency with which detergent is used by binding to and loosening stains. However, the range and specificity of enzymes is very limited today, so they are used in only small amounts. Recovery, regeneration and re-use of the enzymes are also not possible today because these are impractical on a household scale. It is envisaged that, with appropriate R&D, the use and specificity of enzymes will strongly increase and that, with upscaling, enzymes will be regenerated and recycled. This will be possible only in large-scale applications.

Upscaling of the conventional water-based washing process would be facilitated and made more efficient (and therefore also made more attractive to service providers and clients) by switching from batch to continuous washing processes. One suggestion is to develop 'washing tubes' in place of washing machines. Washing tubes would operate on a reverse system. Textiles would pass in one direction down an elongate tube, entering (soiled) at one end and leaving (cleaned) at the other. The textiles would travel in opposite direction to water flowing up the tube. This would allow clean water to be used for rinsing before washing. Detergents and enzymes would enter with soiled textiles at the front of the tube. Upscaling and tube technology would reduce water use and enable waste heat to be recovered.

An alternative to using water as a solvent is to take advantage of the solvent properties of materials in supercritical condition. Supercritical conditions are achieved when the temperature and pressure are such that there is no difference between the liquid and vapour states of a material. If soiled textiles are introduced into a closed environment containing a liquid in critical phase—such as liquid CO_2—dirt will migrate from the item into the medium. In a supercritical state (held under high pressure), the medium behaves and acts as a solvent in place of water. The difference is that it is subsequently possible to separate the waste (the dirt removed from soiled items) from the liquid CO_2 and recover the medium for re-use. This is achieved by changing (lowering) the temperature and pressure conditions. This technology is used today already for the decaffeination of coffee and is interesting because of the facility it offers to recover both the solvent and any cleansing agents.

An important logistical need for the automation of laundry processes and for improved efficiency and performance is for better information about the items to be cleaned. Important information includes the washability of the item, the washing need and the ownership. Some of this information, such as washability, is fixed. Other aspects are variable. Today, garment manufacturers sew labels giving washing instructions (in respect to fixed aspects) into items. Ownership information has to be pinned to individual items by handlers at laundries. Handlers must judge washing need on an item-by-item basis. These turn the sorting and tracking of items into major logistical tasks today. They also make for high labour costs. Most of the sorting and tracking must be done manually. A potential solution is to integrate machine-readable information into textiles, using chip technology and memory; e.g., chips embedded into buttons or memory actually woven into fabrics and fibres. Two stages to this technology can be envisaged: a first stage involving read-only information and a second stage involving self-learning chips and fibres so that relevant information gained over the product life (number of previous washes, for example) can be used to steer cleaning processes. The memory can also be used to store ownership information, which ensures that items cannot be lost.

Box 8.1 **Important technological directions for future textile cleaning**

8.2 includes two new arrangements. One is based on many neighbourhood-scale laundries. The second is based on a few, very large, decentralised laundries. During the research phase, a scenario for each solution was worked out in detail. For each, a development path was envisaged from the present day to 2040, setting out the kinds of innovation achievable or implied by 2005, 2015 and 2040. The scenarios were assessed and refined in respect to their technological, environmental, economic and consumer implications. Consumer questionnaires were used to explore the cultural aspects of each.

Scenario 1: home washing

▷ At present, household members clean textiles themselves using a privately owned washing machine. The washing machine and dryer are designed as stand-alone devices. Each household has its own dedicated machine, giving a 1:1 ratio of washing machines to households. The washing machine and laundry process have been developed for use at the household level. This arrangement calls for unsophisticated technology. To the extent that sophistication is built into equipment, this is to minimise the skill needed by the operator. No expert knowledge is needed

▷ By 2005, the conventional washing process as a stand-alone activity is largely optimised. A start has been made on applying sensor technology and more specialised, stain-specific enzymatic detergents so that washing effort can be better targeted to cleaning needs. Washing machine energy efficiency has been increased by incremental improvement to the motor and insulation.

▷ By 2015, the washing process is more precisely oriented to the requirements of users and to the cleaning need. By the use of more advanced sensor technology and a more complete range of enzymatic detergents, the machine manages the wash process according to whether the cleaning need is intense or light and what kinds of stains are to be removed. Improved enzymatic detergents allow the use of lower wash temperatures. The washing machine is no longer a stand-alone device. Rather it is integrated with other electrical appliances to enable use to be made of waste heat from freezers and refrigerators and it is integrated into the architecture of buildings to allow use of 'grey' water-streams (see Chapter 7).

▷ By 2040, improved sensors and further specific enzymatic detergents have been developed, reducing wastage of washing powders and cleaning effort. Methods for localised cleaning of specific stains together with the development of short wash programmes designed to freshen up textiles rather than fully clean them have reduced the overall demand for textile cleaning. Drying cupboards that use ambient energy and are integrated into building designs are starting to replace conventional electrical drying.

Scenario 2: neighbourhood laundries

▷ Neighbourhood laundries and launderettes exist already in Scandinavia, Germany and Austria but are, so far, underdeveloped in the Netherlands and most other Western European countries. The concept is that of a communal facility located within a neighbourhood or apartment complex. The facility is shared by neighbouring households and can be organised either on a full-service (laundry) or self-service (launderette) basis. The ratio of machines to households is between 1:10 and 1:20. Washing is therefore done in the direct neighbourhood. When operated on a service basis, a semi-professional such as a caretaker who also has other responsibilities within the apartment block performs the cleaning task. This approach makes use of technology of a level similar to (sometimes slightly higher than) that of today's household washing machine. Design emphasis is still on moderating the knowledge and skill requirement of the user. The arrangement requires that textiles are moved short distances between the home and the facility.

▷ By 2005, specialised washing rooms are beginning to be created in some existing apartment blocks and in social housing facilities. The latest technologies for conventional, water-based washing are used—sensors, enzymatic detergents, etc. The high intensity of use of the washing rooms facilitates the recycling of washing powders and water softeners. Also, the latest technologies can be used, such as automatic and detergent-specific dosing systems, multiple component detergents and sensor technology. With more intensive use and a shorter capital replacement cycle, equipment used in neighbourhood laundries has a technological advantage of 5–10 years over households.

▷ By 2015, the neighbourhood laundry service is more commonplace. Facilities are designed into architectural plans for new housing schemes and apartment buildings. Similarly, the service is integrated with several other services such as postal deliveries and waste collection, so that the collection and delivery of textiles is made more efficient. In the washing process, use is made of the grey water-stream. Collected rainwater has several advantages. It is locally available, abundant and soft (i.e. free from dissolved calcium).

▷ By 2040, the neighbourhood washing service is fully developed and accepted. In addition, a more widely defined notion of 'neighbourhood' is applied, extending the concept to streets and localities. Every district now has a neighbourhood laundry. As well as advanced washing routines, enzymes are used to pre-treat stains on a dry basis before washing. Pre-treatment is done at home.

Scenario 3: centralised laundries

▷ Centralised laundry services exist at present, but this arrangement is rather underdeveloped in Western Europe. It is largely restricted to textile-leasing services linked to specific bulk users of textiles such as hospitals, hotels and restaurants. The ratio of machines to the number of users is at least 1:50. Washing is undertaken as a batch process with like textiles being washed together and given like washing treatments. Typically, high temperatures and strong chemicals are used because of health and safety concerns. This means that the technologies used demand rather high levels of operator skill. The processes often reduce textile life and quality. The logistical implications of washing at centralised facilities impose restrictions on the types of textile that are entrusted for professional cleaning. Textiles have to be transported, sorted for washing and potentially resorted after finishing. Clients who lease textiles simply need comparable items to be returned, not necessarily the same items. However, clients who entrust their own textiles to the service want the exact items to be returned. The sorting requirement increases the labour cost component. Because of the high labour costs of offering personalised laundry services and the risks of items going missing or being damaged, centralised laundry services today are mostly restricted to textile-leasing arrangements.

▷ By 2005, laundry services have expanded to take in part of the household need for clean textiles. Small, valuable and personal items are washed at home in conventional machines while bulkier household items are sent to a centralised laundry for cleaning. Laundry washing using water has begun to shift from a batch to a continuous process. Rather than use conventional machines, laundries are beginning to use washing tube technology. The scale of the activity allows for water and heat to be recovered and the washing process optimised to the washing need. Some dry-cleaning (using solvents) is also practised. Textiles are mostly collected from and returned to households as part of the service. Logistics arrangements are still a bottleneck.

▷ By 2015, textile cleaning is more and more professionalised. The laundry arrangement for the bulk wash has allowed the development of specialised technologies for performing the residual small home wash: e.g. small purpose-built machines with a c. 2 kg capacity. Collection and return of the bulk laundry has been rationalised with these functions integrated into already-existing services such as postal delivery, garbage collection and goods delivery. Additionally, households can deliver and collect their textiles themselves using conveniently located collection points in shopping centres and railway stations. The cleaning process at laundries is optimised according to the cleaning needs. Sorting and identifying items is made easier because of 'smart chips' attached to each item.

▷ By 2040, completely new technologies for textile cleaning, drying and finishing are being used at centralised laundries alongside water-based technologies—e.g. cleaning using bacteria and supercritical CO_2. The new technologies enable the cleansing agents to be recovered and re-used. When an alternative to water is used as a cleaning agent, there is no longer need for a drying process, so energy requirements are reduced. The laundry service operators make arrangements for the collection, sorting and return of textiles themselves, not the clients. The smart-chip technology has been further developed and, as a result, damage and loss of items are minimal. Greater security and reduced damage risk mean that, in principle, the total household textile cleaning need can be safely entrusted to centralised laundries.

DEFINITION AND EVALUATION

The development of these scenarios reflects the iterative process of information gathering around the basic ideas mapped out on the solution field and elaborated by participants during the workshop sessions. The process of continuously developing the scenarios and referring them back to participants was carried out in parallel with research into the technological, consumer, environmental and economic aspects of each scenario. Problems relating to technological or consumer acceptance bottlenecks were identified during this process. The concerns raised by one group of stakeholders were often answerable by referring back to others. Consumers' concern over lost or damaged articles in respect to the laundry scenario, for example, raises the question of how such loss might be reduced by using new technology. At the same time, technological opportunities raise questions about acceptability to consumers. In effect, two parallel research agendas emerged: one to address the technological aspects of consumers' concerns and the other to address the consumer acceptability aspects of new technology.

TECHNOLOGICAL ASPECTS

Based on the scenarios and, ultimately, becoming reflected in them, the project identified several promising new technological research directions as complements to already-ongoing R&D on household-scale washing processes. Some of the more important technological implications and opportunities lie in the field of sensors, enzymatic cleansing agents, continuous washing processes, supercritical cleaning processes and chip technology (see Box 8.1).

CULTURAL ASPECTS

The implications of the different scenarios were explored from the perspective of conformity with social values and trends. In particular, time-saving is an important societal trend seen in shifts toward using labour-saving devices, buying ready-made meals and eating out in restaurants (see also Chapters 5 and 6). People are spending less time doing household chores. Another important trend is toward living in higher concentrations in cities and apartment buildings. A third trend is toward smaller—one- or two-person—households. There is an increasing representation of single-person households and households comprising two professional adults without children. All of these trends are consistent with a shift toward the service-product concept.

A questionnaire survey was used to provide structured information about consumers' thoughts and reactions to the scenarios with the aim of identifying criteria underlying declared attitudes and assessing households' willingness to try alternatives to home laundry. Results show that, in principle, the majority of those taking part in the survey had few reservations about using a service provider once an acceptable service is realised. The most important concerns are to do with the quality and convenience of the service and its cost. The quality of textile handling and treatment should be high and the service should exhibit a high level of professionalism and expertise. Results should be as good as those that are achieved at home or better. Items must be returned clean, undamaged and to the correct owner. Liability for lost or damaged items needs to be well regulated and clear. The service must be client-friendly. Opening hours and logistics arrangements should be convenient and the service fairly quick. Those willing to use a service provider suggested a willingness to pay up to US75¢ per kilogram of washing. Almost all households said that there would be some items that they would not want to put out to a laundry service. Although respondents would be pleased to see reductions in the environmental impacts caused by textile cleaning, eco-efficiency was not a major criterion in choosing between alternatives. Changes in behaviour will therefore depend on making laundry services cheaper, better or more convenient than home laundry.

ENVIRONMENTAL ASPECTS

An important research question was whether upscaling could, indeed, offer opportunity for improving eco-efficiency. The technological and logistical arrangements set out in the three scenarios provide a basis for addressing this question, albeit only preliminary.[4] An indicative LCA was therefore performed for each scenario in respect to the three criteria: energy, water and detergent use. Data on the current

4 The details of technologies not yet developed can only be assumed and not precisely measured. Because of this, the analysis was restricted to water-based technologies.

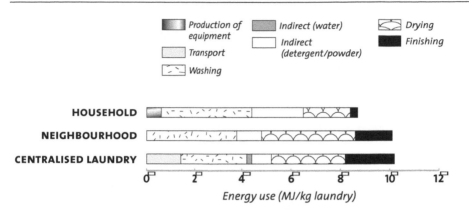

Figure 8.3 **Energy use 1995**

situation provides a benchmark and a basis for target setting (Figs. 8.3–8.6). The analysis was performed for two future dates: 2005 and 2025. Information on achievable environmental improvements was collated from the Group for Efficient Appliances, the Dutch Institute for Cleaning, and the project's industrial participants.

Table 8.4 gives results of the environmental analysis for home washing. Data is given per kilogram of dry weight of laundry both in absolute terms and indexed to the base situation, which is taken to be home washing in 1995. In respect to energy use, the starting situation for home washing is 8.5 MJ per kilogram of laundry (100%). It is expected that this will reduce to 4.9 MJ by 2005 (58% of today's energy use) and to 3.3 MJ by 2025 (39% of today's energy use). The same table also gives

Figure 8.4 **Energy use 1995, 2005, 2025**

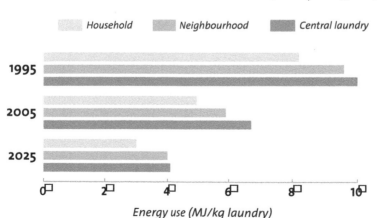

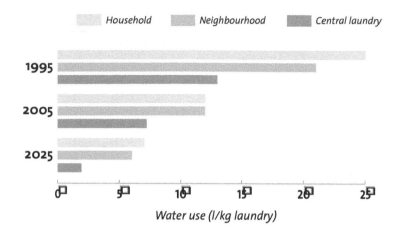

Figure *8.5* **Water use 1995, 2005, 2025**

results for water and detergent use. From a starting position of 25 litres of water per kilogram of laundry in 1995 (100%), water use will fall to 11 litres by 2005 (44%) and to 7 litres by 2025 (24%). From a starting position of 45 grams of detergent per kilogram of laundry in 1995 (100%), detergent use will drop markedly to 22 grams by 2005 (49%) and 16 grams by 2025 (33%). This data represents eco-efficiency improvements of a factor of 2.5 for energy, factor 4 for water and factor 3 for detergent by 2025 in comparison with the base situation.

Several of the anticipated eco-efficiency improvements are the results of synergies between different technological innovations. The improvements in energy

Figure *8.6* **Detergent use 1995, 2005, 2025**

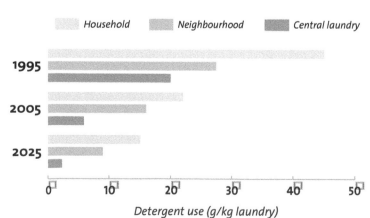

	ENERGY (MJ/kg)	WATER (l/kg)	DETERGENT (g/kg)
1995	**8.5 MJ (100%)** ▷ Wash temperature 55°C ▷ Basic sensor technology ▷ Basic enzymatic detergent	**25 litres (100%)** ▷ Water-level sensors	**45 grams (100%)** ▷ Compact powders ▷ Some enzymes
2005	**4.9 MJ (58%)** ▷ Wash temperature 50°C (better powders) ▷ Improve motor efficiency, optimise insulation ▷ Use less detergent	**11 litres (44%)** ▷ Optimise rinsing ▷ Reduce tub-drum clearance ▷ Recycle water	**22 grams (49%)** ▷ Further development of compact powders ▷ More specific enzymes
2025	**2.9–3.3 MJ (39%)** ▷ Wash temperature 35°C (chemical disinfectants) ▷ Re-use of washing machine components ▷ Improved powders ▷ Use waste heat in households	**5–7 litres (24%)** ▷ Use grey water	**14–16 grams (33%)** ▷ Automatic dosing ▷ Multiple component powders

Table 8.4 **Textile cleaning on a household-scale: technologies and environmental impacts (1995–2025)**

use are achieved mostly as a result of reductions in washing temperature from an average 55°C today to 30°C by 2025 made possible by the introduction of progressively more specific and effective enzymatic detergents and some chemical disinfectants. Energy use will also fall because of reductions in the amount of water and detergent used in the washing process and because of improvement in washing machine efficiency. By 2025, the recovery and re-use of components of old washing machines is also anticipated to reduce the energy input implied by the need to renew the capital stock. Water efficiency improvements are expected to come from the use of sensors to control water levels in domestic washing machines and reductions of tub-drum clearance. Some additional water saving between 2005 and 2025 is expected to come from water recycling and the use of local grey water. The use of detergent will fall because of improvements in enzymatic detergents and the use of automatic sensor-controlled dosing systems.

Table 8.5 gives comparable results for the neighbourhood washing scenario. The starting point for energy use is 9.9 MJ per kilogram of laundry in 1995, which is higher than the use of energy in home washing today (116% of the base situation). The difference is mostly because gas is used for drying in laundries whereas drying at home can partly be done using ambient energy. Energy use will reduce to 5.8 MJ by 2005 (68% of the base situation) and 4.0 MJ by 2025 (44% of the base situation). In contrast, the starting point for water use for neighbourhood washing, 21 litres per kilogram of laundry, is lower than for home washing (86% of the base situation). It will fall further to 11 litres by 2005 (43%) and to 6 litres by 2025 (20%). The starting point for detergent use is 27 grams per kilogram of laundry in 1995 for neighbourhood washing. This is already significantly lower than for home washing (60%). This will fall further to 16 grams by 2005 (36%) and to 10 grams by 2025 (20%). Compared with the base situation, this data represents eco-efficiency improvements of a factor of 2.5 for energy, factor 4 for water and factor 5 for detergent by 2025.

Table 8.5 **Textile cleaning at the neighbourhood-scale: technologies and environmental impacts (1995–2025)**

	ENERGY (MJ/kg)	**WATER (l/kg)**	**DETERGENT (g/kg)**
1995	**9.9 MJ (116%)** ▷ Efficient loads and programmes ▷ Mangles and gas drying ▷ Recirculation of 60% of waste heat	**21 litres (86%)** ▷ Water level control ▷ Efficient loads (large) and programmes (optimised)	**27 grams (60%)** ▷ Expert dosing ▷ Multi-component detergents
2005	**5.8 MJ (68%)** ▷ Optimised apparatus (improved motor efficiency, improved insulation, etc.) ▷ Recirculation of 90% of waste heat	**11 litres (43%)** ▷ Optimal load and water level control ▷ Use of sensors to control the washing process	**16 grams (36%)** ▷ Automatic dosing ▷ More specific components in detergents
2025	**3.6–4.0 MJ (44%)** ▷ Alternative drying techniques, such as microwave drying ▷ Integrated heat- and water-streams	**4–6 litres (20%)** ▷ Use of grey water ▷ Integrated heat- and water-streams	**8–10 grams (33%)** ▷ Highly specific automated dosing using sensors and multi-component detergents ▷ Recovery and regeneration of active agents from waste-water stream

These improvements are achievable in part because of scale economies of resource use when the laundry function is upscaled to the neighbourhood level. The diseconomy of higher energy use caused by the need for artificial drying can be offset relatively soon (within ten years) by recirculating the waste heat from the washing process and using this for drying. Also, energy and water vapour can be recovered from the humid air after the drying process. Substantial benefits can be gained by using the air- and water-streams to integrate the washing and drying processes. Upscaling to the neighbourhood level also allows more efficient use of water. The size of the washing machine can be optimised for larger loads while sensors can be used to regulate the water level within the machine. Grey water can be used for early wash cycles with condensation from dryers used later in the washing process as a high-quality rinsing water. Detergents can be used much more efficiently when the laundry function is upscaled because upscaling allows the use of highly specific and compact enzymatic detergents to be combined with automatic dosing systems and systems for regenerating detergent from the waste-water stream.

Table 8.6 gives findings for textile cleaning in large centralised laundries. The starting point for energy use in 1995 is 10.1 MJ per kilogram of laundry, which is the highest of all three scenarios (118%). The extra energy compared to home washing is because of the need for hot air drying and the requirement to transport textiles over long distances between homes and a centralised laundry. It is esti-mated that energy use will reduce to 6.6 MJ by 2005 (77%) and to 4.2 MJ by 2025 (46%). Today, centralised laundries use 13 litres of water per kilogram of laundry, which is the lowest of all three scenarios (52%). This will fall to 7 litres by 2005 (28%) and 3 litres by 2025 (10%). The starting point for detergent use is 20 grams per kilogram of laundry in 1995, which is also the lowest of all the scenarios (44%). This will fall to 6 grams by 2005 (13%) and 3 grams by 2025 (10%). This data repre-sents anticipated average eco-efficiency improvements of a factor of 2 for energy, factor 10 for water and factor 10 for detergent by 2025 compared with the base situation.

With centralised laundry services, the relatively high energy use at the outset is mostly because of the need to transport textiles. On the basis of assumptions about average travel distances and logistics, the energy used in transporting laundry adds 1.3 MJ per kilogram of laundry to the total energy use, which represents around 10% of the total. Higher washing temperatures and the need to dry large quantities of textiles in a single location also add to the energy use. These initial environmental diseconomies of scale can be reduced and counterbalanced in the future using washing-tube technology and by integrating the washing and drying processes. Because washing-tube technology is a continuous process, it is more efficient in its use of energy, water and detergents. It also offers the possibilities of recovering and re-using resources. Tube technology allows heat and water to be recovered and washing loads to be optimised. The integration of washing and drying processes

	ENERGY (MJ/kg)	WATER (l/kg)	DETERGENT (g/kg)
1995	10.1 MJ (118%) ▷ Need to transport textiles ▷ Need for high washing temperature ▷ Need for gas drying ▷ Tube technology just emerging	13 litres (52%) ▷ Optimisation of loads and water levels	20 grams (44%) ▷ Automatic dosing ▷ Multi-component detergents ▷ Tube technology permits automatic re-use of active agents in the water-stream
2005	6.6 MJ (77%) ▷ Improved mangles ▷ Lower washing temperature ▷ Improved washing tube technology ▷ Integration of washing and drying to recover heat, water and detergents	7 litres (28%) ▷ Filtering and re-use of water ▷ Improved tube technology	6 grams (13%) ▷ Sensors for targeting stains with specific stain removers ▷ Improved range of enzymatic stain removers
2025	3.6–4.2 MJ (46%) ▷ Maintenance of a constant wash temperature ▷ Further integration of washing and drying processes ▷ Use of alternative energy sources, such as district heating and CHP ▷ Improved transport at a systems level	2–3 litres (10%) ▷ Further integration of water-streams ▷ Filtering of water to recover active agents in rinsing water	2–3 grams (10%) ▷ Further improvements in specificity of active ingredients ▷ Active ingredients designed for recoverability and regeneration within the laundry process

Table 8.6 **Textile cleaning on a centralised-scale:
technologies and environmental impacts (1995–2025)**

in laundries will allow condensation from the dryers to be re-used in the washing process. Tube technology, using the counter-current principle, would also enable more efficient use of detergents. The energy requirement for transport will be reduced as logistical arrangements are optimised and as anticipated systems-level improvements in transport technology are implemented, such as a transport system based on electric or hydrogen vehicles (see Chapter 11). In the longer term, energy requirements will be further reduced by the introduction of alternatives to water-based cleaning.

The overall outlook is depicted in Figures 8.4–8.6. These indicate three major conclusions. The first is that the eco-efficiency of all three approaches to textile cleaning can be expected to improve over the coming decades. The least progress is likely to be made on improving the efficiency of energy use and the greatest on the efficiency of water and detergent use. Second is that, at first glance, the position over scale economies is not entirely clear-cut. Upscaling is associated with a slightly higher energy use than home washing, but a substantially lower water and detergent use. These relative differences between home washing and outplacement are maintained even as the absolute eco-efficiency of each approach increases from one time-horizon to the next. Analysis of the structure of energy use across the three scenarios reveals that drying is a major factor in the difference between home washing and outplacement. Centralised laundry services may use three times as much energy for drying compared to home washing. Neighbourhood services may use twice as much. However, a trend is for more households to use their own tumble-dryers, which are substantially less energy-efficient than drying processes within centralised laundries. A higher penetration rate of household dryers (not incorporated into the above analysis) would reduce the energy advantage of home washing over wash services. A third conclusion, considered in more detail below, is that none of the considered arrangements achieves the targeted factor 10-plus eco-efficiency improvements across all three parameters. Improvements range from a factor 3 for energy to a factor 10 for water and detergent.

COST ASPECTS

The cost of washing at home today is around US35¢ per kilogram of laundry. The price paid by users of laundry services is around US$1.90 per kilogram. These figures imply a near six-fold price difference. The cost of using laundry services today is much higher than the US75¢ per kilogram level that the majority of consumers say they would be prepared to pay.[5] Of the cost of laundry services today, labour represents about 45% of the total, taxes and profit a further 25%, capital equipment 15%, transport/distribution 10% and raw materials only 5%. The effects of anticipated technological innovation would reduce the price differential substantially and in a relatively short period. It is expected, for example, that the cost of laundry services could be halved to just less than US$1 per kilogram by efficiency improvements reachable, in principle, by 2005. Given that there would be some efficiency gains and cost reductions achieved also in home washing over this period, the relative cost differential would be reduced by these developments to around a factor of 3. However, this is still too high for laundry services to be attractive for most potential consumers.

5 Even if the household were to employ a helper to do the laundry at home and to pay the current wage rate for home help, home washing would cost US$1.75, which is less than today's cost for professional laundry services.

WRITING 'FUTURE HISTORIES'

At the end of the research phase of the project, one year after the first workshop, the project management team had constructed and evaluated the three scenarios with the help and participation of the many stakeholder and interest groups. As a result of the interactive way that the scenarios had been built, there was a sub-stantial level of consensus among stakeholders that each of the scenarios was an accurate portrayal of a realistic approach to textile cleaning. Also, stakeholders had a high level of confidence in the objectivity of findings and a sense of ownership over the results. The outcome was that stakeholders shared common visions, insights and knowledge about the possibilities and limitations of sustainable textile cleaning. All of the findings were summarised together in a single booklet and circulated as a prelude to the final workshop, which was designed to help partici-pants draw implications from the scenarios so that they could develop and imple-ment an action plan.

This 'installation' phase of the project had three main aspects. The first was to develop a set of work plans. The second was to build new networks of innovators capable of putting the work plan into effect. The networks should develop clearly defined proposals for tackling technological, cultural and structural bottlenecks highlighted during the project. The third was to shift leadership of the project from the STD programme office onto these networks, which would then carry the work forward autonomously. To these ends, the final workshop was organised into two sessions. The first was aimed at drawing out the most significant and important challenges faced in moving toward sustainable textile cleaning. The second was focused on defining lines of research and pilot projects that should be started now in respect to the more important challenges. In order to facilitate the process, the three scenarios and the evaluations of these were translated into the form of a single 'future history'.

The role of the future history is to clarify the co-evolutionary nature of the different kinds of necessary innovation, to propose ways in which bottlenecks might be overcome and to provide insight into the need to work simultaneously on several complementary innovation fronts. Important findings about the possi-bilities and limitations in respect to sustainable textile cleaning can be contextu-alised in the future history in ways that make it easy to visualise their real meaning. The advantage is that the future history presents information in the form of processes operating in a dynamically developing 'systems' context. There are few stand-alone or immutable facts. Nor is there a future to be predicted or forecast. Rather, the future is to be created. With the benefit of the virtual hindsight provided by a future history, stakeholders can decide what they want to achieve and what part they can play in creating the future. They can also make decisions about what they need to do, who they need to work with and what actions they should take first. In effect, the future history is a tool that aids discussion and decision-making about solution directions and supporting measures such as research and pilot projects.

DRAWING OUT THE IMPLICATIONS
FROM FUTURE HISTORIES

It is not possible here to set out a complete future history or to draw out a full set of implications. Rather, several extracts from the future history developed in the course of this project are given as illustrations. The chosen extracts relate to technological and cultural innovations and are presented in Boxes 8.2 and 8.3 respectively. The three extracts given in Box 8.2 are concerned with the logistics of laundry services, the design of new fibres and fabrics and the need to professionalise the textile-cleaning process in order to use solvents other than water. These have been chosen because they illustrate how the future history can be used to develop R&D tasks, timetables and networks. All three extracts hold implications for the need to expand research networks beyond the obvious players and conventional partnerships to build relevant platforms of support for lifetime materials stewardship. The extracts given in Box 8.3 are concerned with the cultural and behavioural challenges, such as the need to address the qualitative aspects of service provision and to focus on niche markets to develop a critical threshold demand for services. An important implication of the future history set out in Box 8.3 is that demand management may be needed as a complement to technological innovation if the environmental stress of textile cleaning is to be reduced to targeted levels.

Several deductions useful for the specification of the challenges posed by sustainable textile cleaning can be drawn from the scenarios and future histories. Upscaling requires enabling technology, especially to facilitate the logistics of handling items and to increase the eco-efficiency of cleaning processes. However, once these fundamental technologies are in place, these could both widen the attractiveness of laundry services to more households and increase the proportion of the total textile-cleaning need that each household would be willing to entrust to laundries. In turn, this would hold further implications for R&D. Positive feedbacks would imply that, as the demand for laundry services increases, the more can be invested in improving the service offered. A stumbling block is the difficulty of generating a threshold demand for laundry services in the near term. Results from the consumer survey show that there is a willingness to use laundry services, but that take-up will depend on consumers deriving tangible advantages from the

66 The problem of clothes going astray was a bottleneck at first. It put many people off using laundries. Also, sometimes, the washing treatments were too harsh and clothes were damaged or aged prematurely. The basic problem was to do with information. Initially, this was tackled by means of better labelling techniques, but this was not very effective. Then chips were used. At first, these were attached by the laundry to the items to be washed and were used to store information about the washing process and the owner. The chips allowed more automation of handling and washing, reducing the risk of damage and loss. The costs of the service were also reduced because the chips did away with much of the manual sorting and handling. Now we also have memory that is woven into textiles during manufacturing and is invisible to the owner. This means that the textile industry is also now taking greater life-long responsibility for fabrics and, as part of this, for any mistakes in the washing process. This has led to much higher-quality textiles. 99

66 Much progress has been made in the area of fibres and fabrics. In the hundred years following the first development of synthetic fibres and fabrics, many different types of fibre were synthesised to improve performance on traditionally important criteria such as strength: weight and cost:strength ratios. In addition, by the end of the 20th century, we already had fibres and fabrics that were optimised for specialist uses: for example, strong, light, absorbent and insulating fibres for sportswear, and fire-resistant and bullet-proof fabrics for use by the security professions. The success in building properties such as these into fibres and fabrics was an indication of what could be achieved once a specific property is perceived as important and is specified as a design criterion. In recent years, the cleaning properties of fibres and fabrics (and the environmental implications of fabric cleaning specifically) have become much more important. Synthetic fibres and fibre treatments have been developed so that these are dirt-resistant and dirt-shedding, have low water absorbency, and are non-creasing. These fibres and fabrics now have a big share of the textile and clothing market. 99

66 The use of water and detergents for cleaning clothes has strongly decreased, but reducing the energy needed for cleaning textiles has been a long-standing bottleneck. The reason is the use of water in the washing process. Water has to be heated to be an effective solvent for dirt and stains but it must then be removed by evaporation in the drying process. Many years of research went into finding alternative cleaning processes based on solvents other than water. Dry-cleaning using organic solvent is much less energy-intensive and the solvent can be recovered. However, this process—like other non-water processes—is only viable when the cleaning task is professionalised. It is not suitable for use within households. Also, some textiles and dyes are not compatible with organic solvents. Close liaison has been needed between the laundry service providers and the manufacturers of textiles and dyes to solve this problem. 99

Box 8.2 **Extracts from future histories of technological innovation**

Flexibility for the consumer is an important aspect and was a bottleneck in the beginning when clients had to be at home for the wash to be collected and returned. This was inconvenient and inflexible. To be successful, the service needed to become consumer-friendly and to offer a choice of pick-up and delivery arrangements without the need for the client to be present. In suburbs, wash-boxes and bins began to be used in which soiled or cleaned textiles could be deposited, while in urban areas centralised wash containers were placed where soiled laundry could be deposited. The logistics of clean laundry delivery in cities became integrated with several other services (postal services, goods delivery, etc.) and special arrangements were made in many apartment buildings with a room being set aside for such arrangements.

The growth in the number of professional couples without children has continued and the trend is for this group to spend less time doing household chores themselves. This group had a special interest in putting the washing out to a professional service provider and provided the first strategic niche market for improving laundry services. With some early success the growing market made for greater competition among service providers, through which prices were reduced and innovations were encouraged. As a result, new target groups could be enlisted as clients. The list now includes more old people, single-person households and people living in small apartments.

Through the years it became apparent that consumers do not want to entrust all the wash to a laundry service. Especially, they want to do a small amount of personal washing at home in their own washing machines with the rest given over to a professional wash service. For the personal wash, very small machines have been developed. Especially in the beginning, this arrangement also suited the service providers and was encouraged by them. Until the arrangements for handling and cleaning items were perfected, there was a greater risk of small and expensive items being lost or damaged. It was better in the beginning that laundries were not responsible for the most critical items. Later, as logistics arrangements and washing processes were perfected, consumers and service providers became more confident in the service and a bigger fraction of the total wash is now put out to service providers.

Through the years all kinds of additional services have been introduced by the laundry operators. Textile cleaning is now a diverse market with substantial growth in both products and services. Altogether, the choice available to the consumer has become much bigger. Nowadays, stress is placed on high quality. The service is grounded in professionalism and has a higher status and esteem than laundries in former times. Being environmentally friendly is seen as part of the quality aspect of laundry service by consumers and is part of their attraction.

The environmental impacts of textile cleaning are being reduced by means of active innovation, but even today's 'greenest' washing processes cannot reach the sustainability targets formulated in the 1990s by the STD programme. Only laundry services can reach factor 10-plus reductions in respect to detergent and water, but even these cannot reach the targets for energy. So we needed additional directions, especially to reduce the demand for textile cleaning. The quantity of washing per capita was increasing in the 1990s and early years of the 21st century. Research has been established to look at reasons for the increase and at ways of reducing the amount of material that must be washed.

Box 8.3 **Extracts from future histories of cultural innovation**

service. Quality standards must be high. The service must be convenient and reliable. Responsibility for losses and damages must be well regulated. Prices must be substantially lower than they are today and the quality of the service and the cleaning results substantially higher. The price gap between the willingness to pay for laundry service (up to twice home laundry costs) and current laundry prices (five to six times current home laundry costs) is too great at present. Much of this price gap is made up of costs that are amenable to policy influence, such as the direct and indirect tax cost components.

The environmental benefits from upscaling mostly lie in the possibilities for recovering and re-using energy and materials and for efficiently matching cleaning treatment to need so that fewer resources are committed to the process for each kilo of laundry in the first place. While a centralised water-based textile cleaning service is more eco-efficient than home washing in respect to water and detergent use, energy use is still a stumbling block with any water-based process. If there are immutable limits to the improvement of water-based technologies, this suggests that technologies based on chemical solvents may be needed. These are only feasible if provided at centralised facilities. Moreover, the demand to develop the relevant technologies can only come from large service providers who, if they have a sufficient client base, will be driven in this direction by competitive pressures to cut costs. Both of these imply that it is important to develop water-based centralised washing services in the short to medium terms as a necessary stepping stone to more eco-efficient non-water-based technologies in the longer term. A substantial additional reduction in environmental impact could come from demand reduction. The environmental analyses indicate that there are immutable limits to the eco-efficiency of water-based textile-cleaning technology. Until alternative technologies become available, demand reduction is the surest way of reducing the absolute environmental burden of textile cleaning. This suggests the need for research into the cultural and ritualistic aspects of washing. Reducing environmental impacts implies the need for a closer co-operation between all those involved in the cradle-to-grave management and use of textiles. This group includes the manufacturers and operators of textile-cleaning technologies, all those involved in the manufacture of textiles, and consumers.

CONCLUSIONS

The laundry case is important in its own right because of the environmental burden that textile cleaning imposes and because of the limited scope for reducing this burden by incrementally improving the domestic-scale washing process. An alternative innovation trajectory based on moving the laundry function into dedicated laundries would allow new process technologies with lower environmental impacts to be used to clean domestic textiles in the longer term. Environmental benefits could potentially arise because of the greater scope that upscaling offers for technological and operational improvements in laundry processes. The major bottleneck is that the larger-scale technologies with the greatest eco-efficiency potential are unlikely to be developed unless a demand for centralised laundry services emerges in the short to medium term. This is not easy to achieve as it will involve significant changes in behaviour without bringing immediate advantages to service users. The task thus becomes one of starting a new innovation trajectory in the short to medium term that offers long-term scope for improving environmental efficiency. The future history offers a means of demonstrating this bottleneck and of exploring its implications so that these can be translated into short-term action steps.

The most important tangible outcomes from the STD laundry project include an R&D agenda for the development of enabling technologies that has been taken up by a new innovation network, an innovation network that is looking at sustainable textiles and a neighbourhood-scale pilot project that is experimenting with washing services. Within the context of the pilot project, research will be carried out to explore the cultural and behavioural aspects of washing. In addition, the STD laundry case study holds wider implications because of parallels with other unsustainable household activities. The issue of whether there are opportunities more generally to reap *eco*nomies of scale is now being researched within the framework of a much bigger European project on sustainable households. This is exploring other opportunities to substitute household-scale machinery with service products and is evaluating potential eco-efficiency benefits. The research is a spin-off from the STD project and is using similar investigative methods.

Chapter 9
Redirecting R&D resources: the case of chemicals and industrial materials

This chapter and the two that follow are closely related. The present chapter presents an overview of the STD programme's case study of the Dutch petrochemicals, bulk chemicals and fine chemicals industries. In it we discuss the method used by the STD Chemistry Project to move toward a consensus on strategic technological directions to follow toward reaching sustainability. The current chapter presents a sector-wide review of the main 'source-to-service' material chains, asking how these might be reconfigured to be more sustainable. Since the material chains cut across sectoral boundaries, their reconfiguration holds implications both within and beyond the chemicals sector. In the following two chapters, the focus is on the results and outcomes, in the form of evaluations of particular solution directions and of how well these are now embedded in active R&D efforts in the Netherlands.[1] All three chapters are aimed at demonstrating the need for strategic-level analysis to identify sustainability challenges, set up appropriate innovation networks and target R&D resources effectively. An important conclusion is that modelling 'source-to-service' chains forms an appropriate basis for reviewing sources of unsustainability in present chains and for proposing and evaluating technological alternatives based on forging a connection between renewable raw materials and sustainable products.

The Dutch chemicals sector is large, very important to the economy and makes major claims on eco-capacity. It is widely understood within the industry that many of today's industry practices are incompatible with long-run sustainability and that change is needed to meet sustainability challenges. The sector is also a

1 The two later chapters provide progressively greater depth and detail. In turn, they cover a sub-sector, bulk organic chemicals, and a specific need-based source-to-service chain, mobility.

key repository of the technical expertise needed for developing sustainable technologies in response to its own sustainability challenges. However, the same R&D resources are needed also for meeting sustainability challenges in other sectors of the economy. For example, the chemicals sector could play a major role in developing the technologies that would enable a future sustainable society to make better use of solar energy (photovoltaic cells), salt-water (membrane technologies) and currently unproductive regions such as arid and saline soils (gene technologies, bio-technologies). So it is important that the R&D resources of the sector are appropriately and strategically directed. The major barriers to this include the difficulties of achieving cross-industry and cross-sector consensus on long-term strategic directions, the commercial pressures to focus R&D resources on new product development for today's markets and the adverse market conditions for sustainable solutions, which are not yet cost-competitive.

THE ROLE AND CURRENT STATUS
OF THE CHEMICALS SECTOR

The role of the chemicals sector is to process raw materials drawn from nature and to add value to them by increasing their usefulness as fuels or as products with specific sets of energetic, structural or functional properties. The value-adding process can be represented as a source-to-service materials chain that begins with raw materials extracted from nature and ends with the final products that are used to deliver needed functions. The sector is extremely large and diverse. For convenience, it can be subdivided on the basis of the scale of production (bulk versus fine chemicals), on the nature of the materials being transformed (organic, inorganic), on the nature of its products or of the needs that its products meet. Irrespective of these divisions, the basic role of all branches of the sector is to produce products with desired property combinations from raw materials drawn from nature. Products with desired performance characteristics are identified together with a feasible production route involving one or more feedstocks and a set of conversion, separation and reformation stages. A wide menu is available from which to select conversion and separation combinations for producing and upgrading the product stream. Production facilities to realise the production are then designed, constructed, put into operation and optimised. For illustration, Table 9.1 presents a set of products and the needs that they are designed to fulfil. Table 9.2 provides examples of conversion processes.

The sector can thus be considered as the sum of a whole set of interrelated and partially integrated materials chains. The key elements of the sector are the basic raw material resources drawn directly or indirectly from nature, the set of starting compounds developed from these and the set of intermediary and final products

NEED TO BE FULFILLED	CATEGORY OF CHEMICAL PRODUCT
▷ Protection against corrosion/wear	Coatings
▷ Provide colours to specification	Paints/pigments/dyes
▷ Clean clothes/materials	Detergents
▷ Suppress undesirable biological agents	Pesticides/herbicides
▷ Connect separate parts	Adhesives
▷ Lubricate machinery	Lubricants
▷ Provide materials for textiles	Fibres
▷ Maintain health	Pharmaceuticals
▷ Provide packaging/construction materials	Bulk polymers

Table 9.1 **Some chemical products and the needs they are designed to fulfil**

produced by the sector. In addition, there are the sets of reaction sequences and processes used to transform the materials. These involve additional materials such as catalysts and also the specialised equipment needed to provide appropriate reaction conditions and to control reactions. As well as producing desired products, most reactions produce by-products and wastes. Necessarily, as one of the major intermediaries in the life-cycle chains of materials, the sector is environmentally very relevant. It is also a major repository of the scientific knowledge and expertise necessary for lowering environmental burdens both in this and in other sectors.

A modern economy and society capable of meeting accustomed needs would not be possible without using materials. Society and the economy need materials for structural and functional purposes and also as energy carriers. Even in a very 'dematerialised' future there will still be a need for functions that can be delivered only through products that have a material basis. By implication, there will always be a need for chemical and material products and, therefore, for ways of producing

Table 9.2 **Some important conversion processes**

▷ **Oxidation**	▷ **Reduction**
▷ **Hydrogenation**	▷ **Reforming**
▷ **Chlorination**	▷ **Cracking**
▷ **Formylation**	▷ **Electrolysis**
▷ **Esterification**	▷ **Hydrolysis**
▷ **Polymerisation**	▷ **Pyrolysis**

these sustainably. Nonetheless, by virtue of the very nature of its operations, the chemicals industry is environmentally relevant. It mobilises and processes large quantities of raw materials that are extracted from nature as primary products of mining, agriculture and forestry. Many of these materials are non-renewable. Others are toxic or hazardous themselves or are associated with toxic or hazardous co-products. Some chemical processes produce products that do not occur in nature, have no natural breakdown routes, are toxic or have undesirable environmental effects. In order to maintain the positive contributions of the sector in the fields of income generation, employment and contribution to human welfare, and to make these contributions compatible with sustainability, the industry will need to be transformed. If a sustainable form of chemistry is to be achieved in the future, government, research institutes and industry need to work together now to develop a strategic vision and to initiate long-term research agendas for technology renewal.

DYNAMICS AND DRIVERS OF CHANGE IN THE SECTOR

Change is nothing new in this sector. In many ways it has been changing ever since it began commercial-scale operations in the 18th century. Apart from competition, a continuous pressure for change that has strengthened recently because of globalisation, the major drivers have been new scientific discoveries,[2] changes in the price or availability of raw materials and changes in the regulatory and market context. These have led to changes in the fields of resources, processes, products and markets and to some fundamental transitions in the industry. One of the most significant has been the shift in the raw materials basis for organic chemicals over the last century, first to coal, then to oil, and more recently to natural gas. When coal was the main raw material source, acetylene and aromatics were the main starting compounds in the production chain of organic chemicals. Whole families of organic chemicals were derived from them. When coal gave way to oil, ethylene and benzene replaced acetylene as starting compounds. Later still there was a shift to naphtha. Another important long-term trend has been the development of synthetic organic products as alternatives to materials either found in nature or derived from renewable raw materials.[3] There have also been phase-outs of some products with known health or environmental effects, such as DDT and CFCs.

---

2   Important discoveries have been made concerning new molecules, whole new categories of material products, new properties of materials, and new forms of synthesis or catalysis.
3   Examples include synthetic rubber and fibre. The first synthetic fibres—viscose rayon and cellulose acetate—were made at the end of the 19th century. Dozens more have since joined them and R&D on synthetic fibres continues strongly today.

The industry has also grown substantially. Because its products typically change hands at point-of-sale—which, in the context of diminishing marginal production costs, means that additional sales add to profit—the sector is strongly output-oriented. At the same time, there has been a diversification and proliferation of production processes and products. As a result of emerging new needs for materials with specific properties, especially in respect to growth sectors such as micro-electronics, there is a continuous creation of new classes of product, such as for electricity-conducting polymers. It is estimated that, on average, six new chemical products are brought into commercial production each day—equivalent to approximately two thousand each year. The chemical products of the industry have become increasingly specialised and there has been an explosion in product diversity. In the process of developing new products and production routes, highly sophisticated, multi-stage conversions and separations have been introduced. Due to the demand for more complex products and for product diversification, the number of unit operations in production processes has increased along with the complexity of the connections among these. Yet, because of competitive pressures, the time available for developing new products is shortening. Monitoring and control have therefore become ever more important aspects of chemical industry operations.

The R&D agendas of the chemicals sector are also bound up in these drivers and processes of change. One important aspect is that R&D efforts have increasingly become focused where most value can be added. In value-adding terms, the 'centre of gravity' of the market has shifted downstream to high-technology growth sectors such as aerospace, micro-electronics and healthcare. The R&D emphasis of the materials and chemicals sector has shifted accordingly. It has moved down the material chain to focus on products and, in respect to products, across from bulk products to higher-value fine chemicals. Since the Second World War, R&D interest has moved to polymers, ultra-light composites and speciality materials for limited applications, such as semiconductors. Increasingly, the focus of R&D in respect to these markets is on developing chemical products with sets of precisely speci-fied—customised—property combinations.[4] By implication, R&D activities are more specialised than ever, focusing in detail on the properties of specific materials destined for highly specific applications in particular markets.[5]

4 This largely explains the post-war success and growth of plastics since property combinations are more easily customised with plastics than with more traditional materials such as metals or ceramics.

5 An important point well illustrated by developments in the area of synthetic fibres is that R&D has generally been concerned less with advancing any single property and more with developing materials with customised property combinations. Many of the most recent synthetic fibres perform less well on established basic properties than earlier synthetic fibres. Their commercial attraction is that they offer desirable property combinations (Rohatgi *et al.* 1998).

In recent years, pollution control has also entered the R&D equation. Research efforts in this regard have been effective.[6] Nevertheless, environmental R&D have been largely limited to 'good housekeeping' and 'incremental improvement' in respect to existing production processes. Although essential, these efforts are insufficient to shift the chemicals and industrial materials sector toward sustainability. While end-of-pipe approaches allow wastes to be captured and treated before being released into the environment they neither reduce the draw on non-renewable resources nor the overall quantity of wastes arising. Most often, they actually increase both, as more raw materials have to be brought into the economy to build extra equipment and operate additional processes. Moreover, much of the positive environmental effects of actual or anticipated technological contributions from these efforts have been or can be expected to be cancelled out by growth in the production and consumption of chemical products. Above all, end-of-pipe responses have no effect on the quantity or type of raw materials used by the sector and so will not deliver long-term sustainability.

KEY SUSTAINABILITY CHALLENGES

We use materials to provide structural, functional and energetic services in a wide range of application areas, making use of the physical and chemical properties of materials found directly in nature and of the possibilities of engineering desirable and useful combinations of properties into materials made by chemical synthesis. The demand for chemicals and materials is increasing and, also, there is pressure to provide increasingly sophisticated products with specialised combinations of properties and even 'built-in' intelligence. Such products tend to involve many process steps and the production processes are often materials-, energy- and waste-intensive. R&D in the chemicals sector focuses heavily on developing new products to meet high-value demands. However, from the perspective of sustainability, the long-term challenges for the sector lie in a combination of dematerialisation of the wider economy (using fewer materials to provide the same or better functionality) and in developing more sustainable source-to-service material and energy chains. Becoming more sustainable means reducing the overall annual draw on non-renewable resources, switching to renewable resources, using all materials and energy more productively, avoiding the dissipation of toxic and non-biodegradable

---

6   In 1994 it was reported by the Dutch Chemicals Industries (VNCI) that an environmental care system was operational in 140 production locations in the Netherlands. Reported emissions of phosphate to water have been reduced by 73% from 1988 to 1992 and of heavy metals by 60% from 1989 to 1992. VNCI and the government have signed a covenant aimed at realising Integrated Environmental Targets by 2000 and 2010.

materials and closing materials cycles. The use of some toxic substances may need to be substantially reduced, restricted to certain applications or phased out altogether.[7]

For the bulk organic chemicals industry, the major challenge is to forge a connection between the domain of renewable raw material resources on the one hand and sustainable products on the other. The need is for versatile starting compounds that can be made from renewable feedstocks and used to make a broad spectrum of products. Related to this is the need to reorganise downstream production processes so that these can make use of feedstocks produced from renewable raw materials. In effect, this involves closure of the materials cycle through nature. For the inorganic sector, post-consumption wastes will need to be recovered and recycled through the economy. This is relatively straightforward in the case of bulk materials used in large structural applications. The problem arises when dealing with smaller quantities and mixtures of inorganic chemicals in solid, liquid or gaseous matrices, such as spent cleaning or cooling materials, where recovery of individual constituents is more difficult. In order to realise this option, new, drastically improved recovery, separation and upgrading technologies will be needed with downstream production processes reorganised to make use of recovered materials.[8] For the fine chemicals industry, the challenge is to increase the efficiency of conversions substantially, because, at present, these are inefficient and produce large amounts of waste. In addition, sustainability creates wholly new demands, such as for 'lightweight' materials with improved performance relative to their mass. It also creates new needs—such as to capture solar energy, which heightens interest in the photoelectric properties of materials and similar phenomena. Possible approaches for dealing with these challenges can be summarised as a decision chart (Fig. 9.1).

Shifting toward sustainability presents a strategic challenge for the chemicals sector that will require a reorientation of innovation effort. Many of the needed changes imply technological innovations either at the front end of the materials cycle in respect to raw material sources and primary conversion processes, or at the end of the materials chain, where innovations are needed to close the loop through recycling or re-use of post-consumption wastes. Nonetheless, the bulk of the sector's R&D resources are still devoted to product and process improvement

---

7   Especially, there is a need to reduce the use of heavy metals such as arsenic, cadmium, chromium, lead and mercury, phasing them out in applications where use is inherently dissipative, such as in dyes, paints, coatings, preserving materials and allowing them to be used only when the materials can be fully recovered and on the basis of detailed schemes for recovery. The use of halogenated hydrocarbons may also need to be restricted.

8   There are several technological opportunities. Candidates at the level of molecular separation include those based on chromatographic and supercritical techniques. At the level of particles, mechanical separation is possible after individual particle recognition, which depends on developments in the field of advanced sensors and control systems.

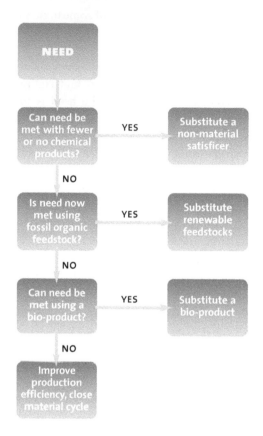

The chemicals and materials sector can be split into broad categories, especially along the lines of organic versus inorganic products and bulk versus fine or speciality products. In the case of some needs now met using chemicals, it might be possible to meet the need differently in a sustainable future: for example, without using chemical products at all or by substituting alternative chemical products that offer higher resource productivities than the ones used today. In the case of needs still met using chemical and material products, the options for a sustainable future are either to provide these on the basis of renewable raw materials or to close the materials cycles. The main product categories that are difficult, if not impossible, to substitute—albeit that levels of demand may change—are fuels, organic chemicals and structural materials used for the fabrication of products and for construction. The first two of these categories can be met from renewable sources of energy and materials. There is the possibility, also, of substituting products of organic chemistry (bio-products) for inorganic products in some structural applications and in some other areas of need. In the case of needs that continue to be served by products of inorganic chemistry, the major issues concern improving the efficiency of production processes and closing the materials cycles.

Figure 9.1 **Possible actions toward sustainability in the chemicals and materials sector**

in the middle sections of the materials chain.[9] Thus, there is a need to shift the R&D agenda and research resources toward more fundamental issues of technology renewal. There are other obstacles to be overcome, too. Because the material chains cross industry and sector boundaries, there is need for relevant actors to come together to agree on strategic solution directions. Moreover, concerns over the environmental sustainability of the sector are embedded in other imperatives. Production routes are getting longer and products are becoming more complex, while the time taken to go from product design to marketing is getting shorter. This adds to the problems of simultaneously optimising production routes, meeting health and safety requirements and maintaining competitiveness. The sustainability challenge to the sector needs to be contextualised in this wider, dynamic context, since meeting the challenge requires solutions to be found that can address these other challenges simultaneously.

STD METHODOLOGICAL APPROACH

The STD Chemistry Project began with a stakeholder analysis to establish the interests and perspectives of the key players. In the Netherlands three major international firms—Akzo-Nobel, Shell and DSM—account for 80% by value of total chemical-sector production. Other key actors are potential suppliers of renewable raw materials for the sector, industrial consumers with interests in the sector's products and those with interests in sustainable technologies that might be developed within the sector. Three such sectors are agriculture, transport and electricity generation. A central STD concern was to engage the active enthusiasm of actors from all these sectors to work together toward long-term sustainable solutions capable of meeting the aspirations both of society and industry. Another concern was to concentrate STD efforts where the involvement would add significantly. A set of high-level meetings was arranged and an 'industry club' set up to discuss stakeholders' potential interests in a case study. The work programme was co-developed and co-financed by the STD programme and members of the industry club.

It was decided that a group of outside experts from universities and institutes of technology should be invited to develop and present their visions of long-term

---

9   The majority of R&D expenditure today is devoted to developing new products and improving existing products. In contrast, there is relatively little emphasis either on broader strategic issues confronting the sector or on questions concerning the surety of basic raw materials supply at the front end of the materials chain. The implicit logic that underlies this rests on the assumption that the raw material source is a factor neither in continuity of production nor in future competitiveness among producers.

possibilities for the sector. These would be developed in relation to five illustrative need-based source-to-service chains chosen to represent the different kinds of sustainability challenges facing the sector. In this way, the presented visions would not represent the position of any single stakeholder, nor would any visions be ruled out from the start. The ideas put forward in the essays were presented, critiqued and improved at a forum before an invited audience including representatives of the chemicals industry, government bodies, technology institutes and societal stakeholder groups.[10] The purpose was to identify potentially cross-cutting and synergistic technological themes as the basis for starting new innovation pathways and for accelerating promising pathways already under way. The forum was used as a way of breaking down barriers between actors and initiating new networks capable of advancing the chosen pathways. This way of approaching the issues gave all the key actors an opportunity to be involved in the search process and to have common and simultaneous access to new perspectives and ideas brought forward by the outside specialists. At the same time, it provided opportunity for discussions generally precluded by usual business practice and concerns.

An important consideration was for the essay topics to span, as fully as possible, the range of challenges and opportunities facing the sector. It was envisaged that, by working on several challenges at the same time, the work might lead toward a coherent, strategic vision for the future development of the whole sector, which would include the development of mutually supporting clusters of technologies. Three of the essay topics were chosen so that the essayists could demonstrate how a need currently met by the chemistry sector using unsustainable technologies could in principle be met using sustainable technologies. A further two explored how expertise within the chemistry sector might contribute toward meeting sustainability challenges in other sectors. The focus was on key areas of need where sustainable technologies would bring the greatest quantitative benefits. The essays and the reasons for choosing them are listed below.

▷ **New resources for the chemical engineering industry.** This was chosen because of the current heavy reliance of the bulk and fine chemicals

10   As an exercise in 'learning-by-doing', the STD programme continuously evaluated and refined its methodology. It was found from earlier STD projects that creativity workshops, although useful instruments for tabling new ideas for innovation, need to be focused and structured to be successful. Especially, some of the early workshops had led to some confusion between tabling and critiquing ideas. To impose focus and structure in the case of the Chemistry Project, as well as to obtain a set of genuinely high-quality insights, it was decided to use an essay approach that would allow creativity and evaluation to be separated through a division of roles. This was needed especially in this case because of the highly technical nature of the subject matter, which meant that the community of specialists and experts had a central role to play in both proposing and evaluating new ideas. The essayists, working in pairs to provide initial feedback on ideas, presented their visions about future technologies. These were then subject to peer review and to critique by interested parties at a forum.

industries on fossil carbon as a source of raw materials and energy and on other non-renewable raw materials. A priority is to find renewable production routes for major bulk products that nowadays are made from non-renewable raw materials—especially for liquid fuels, bulk organic chemicals, polymers and structural materials.

▷ **Agricultural production in brackish soil.** Soil salinification was selected as a problem for which chemistry is not responsible but where its expertise is relevant in finding solutions. Soil salinification is a growing problem for food production but also has implications for the capacity of agriculture to supply materials and energy to the chemicals sector in the future on a renewable basis.

▷ **Light materials for support structures.** Structural materials were selected for review because strong and lightweight materials made from renewable sources are needed for applications, such as vehicle construction, where synergies between weight reduction and energy saving could be realised through dematerialisation.

▷ **Colouring textiles without using dyes.** Colour chemistry was chosen because dyes and pigments constitute a large product category and because currently used dyeing technologies are among the most unsustainable within the entire chemicals sector. Today's dyeing technologies combine the use of non-renewable raw materials with the generation and dissipation of toxic wastes both during production and during disposal of post-consumption wastes.

▷ **New reaction media for precision chemistry.** It is important to develop leaner and cleaner production processes or dematerialised alternatives for fine chemicals, since today's production processes are highly inefficient and produce large quantities of wastes that are difficult to handle.

Each expert (two per assignment) was given a similar brief in relation to his/her topic. Each was given the main boundary conditions within which a vision of a sustainable future should be created. The end-service needs that the sector meets today and any anticipated expansion in these should be met in the future. The sector would need to work within other accepted boundary conditions: for example, those concerning health and safety. In addition, the sector should work within environmental constraints. The essayists were to assume that, by 2040, no fossil carbon resources would be available to the sector and that the productivity of inorganic materials should be a factor 20 higher than today. Would it be possible to sketch an attainable vision of a sustainable future that would meet these conditions? What source-to-service chains and what needs for new source materials, technologies and logistics arrangements would this future vision imply? Can the needs for new technologies be articulated clearly and prospective synergies among possible technological approaches identified?

In the context of the different essays, each of these broad questions was made more specific. If oil and gas are no longer available to be used by the chemical engineering industry, what alternative raw materials, chemical processes and products might be substituted to provide power, fuel, fibres and plastics or to meet the needs that these products fulfil today? If productivity gains of a factor 20 are to be made, how might the efficiency of precision chemistry be improved and materials cycles closed within the economy? How might chemistry provide ways of fulfilling important functions, such as the colouring of textiles, if conventional pigments and dyes whose use is inherently dissipative are no longer to be used? In each case, as well as setting out a vision of the source-to-service chains of the future, the essayists were required to write a future history describing the development of indicated technologies in terms of breakthroughs needed and by when. Essayists were asked to describe the breakthroughs in terms that could be translated into a research or design brief. They were also asked to indicate the status of ongoing applied research on related technologies or of ongoing basic research on phenomena—such as bio-catalysis—that could be explored for practical applications in the search for sustainable technologies of the future.

For illustrative purposes, the key lines of argumentation in respect to two of the essays (colouring textiles and precision chemistry) are given in Boxes 9.1 and 9.2. The content of the other essays and the research lines developed from them are covered in more detail in other chapters.

RESULTS AND OUTCOMES

The essays revealed a number of research themes that were taken up first as illustrations within the STD Chemistry Project but have since become firmly embedded within the framework of ongoing commercial R&D activity in the Netherlands. These include work on C_1 chemistry (reported more fully in the next chapter), on the development of organic photovoltaic foils, on the production and extraction of useful chemicals from plants, on the development of bio-composites and on specific processes used in fine chemistry. There are close synergies between these fields of research, especially between the first four because each is party to the meta-goal of forging a link between renewable resources and sustainable products.

In the field of C_1 chemistry, a strong impetus has been given to accelerate and expand research on biomass conversion with a view to realising large-scale methanol and bio-crude production. This includes work on biomass gasification,

TODAY'S CHEMICAL DYES AND FIXING AGENTS GENERATE TOXIC WASTES THAT ARE dissipated into the environment via the waste-water streams from dye shops. Over 70% of the total contamination generated by the textile sector is generated during the dyeing process, which is often carried out in small-scale and decentralised facilities. Today's dyeing technologies include both batch and continuous processes. The main environmental damages are associated with waste-water contamination by heavy metals, pigments, salts, solvents and additives used during the dyeing and fixing processes. About 1,000 tonnes of dyestuff is used today in the Netherlands, of which 200 tonnes is lost to the environment. The key problems with present-day technologies are that the efficiency of making use of dyeing materials is low, so that proportionately large waste-streams are generated, while the generated wastes, which are toxic, are dispersed into waste-water streams from which they are difficult to remove.

A succession of technological innovations is envisaged that could reduce and, ultimately, solve these problems. Each technology builds on the earlier innovations. The first step would be to dye textiles using contactless methods; possibilities include ink-jet printing for patterns and supercritical solvent technologies for block colours. A second step is to develop useful chromophores, which are complex organic compounds with a double C–C structure and an excitable electron. The electron can be moved under the one-time influence of a laser beam. After a short time it falls back to its original position, but thereafter re-radiates light at a specific wavelength, which gives the impression of a specific colour under the influence of sunlight. (Coloration in nature is achieved through this phenomenon and it is the principle, also, behind the organic photovoltaic cell.) A sequence of chromophore technologies can be envisaged beginning with single-colour (monotonal) chromophores, then chromophores offering multiple-colours (multitonal) chosen and set by laser beams of different frequency and, finally, reversible chromophores. Reversible chromophores would be polymer compounds that could first be fixed to any desirable colour using laser beams in specific frequencies and then reset using electro-magnetic fields, thus enabling colours and patterns to be changed. Research is needed to produce a group of reversible chromophores that can be applied to fibres using ink-jet technology and which can be activated to any desired colour and pattern combination using programmable lasers (Duijsens and van der Schuyt 1996).

The sustainability advantages of contactless colour encryption by laser beams in comparison with dyeing and printing would come from the avoidance of solvents and heavy metals, the low loss of chromophores, the flexibility for small production runs (which reduces textile wastage in the colouring process) and process reversibility. As well as avoidance of water contamination in the process, its reversibility would mean that more textile fibres could be recycled.

Box 9.1 **Colouring textiles without dyes**

hydro-thermal upgrading and flash pyrolysis. A field of research that emerged in conjunction with this is to accelerate work on low-cost and efficient organic photovoltaic foils that convert sunlight into electricity but which might eventually be used also in the production of hydrogen gas for enriching the products of biomass gasification. In the field of direct extraction of useful molecules from plants, experiments are under way using hemp as a source of fibres, high-grade cellulose and oil, and using the potato as a source for C_5 and C_6 chemistry. Both species grow well in Dutch conditions. In the long run it is expected that knowledge gained by working to improve the quantity, quality and separability of the useful fraction, such as by genetic modification, would be transferable to other crop

IN THE FIELD OF FINE CHEMISTRY, THE MAJOR ISSUE IS TO ACHIEVE MORE EFFICIENT production processes with higher yields and lower quantities of problematic waste. This would lead to both dematerialisation and less environmental stress. Today only a fraction of the raw materials used in the production of fine chemicals is converted into final products. Typically, more than 90% of the starting compounds and auxiliary materials used to control reactions and to buffer product streams end up as wastes. This is due largely to production arrangements. Small quantities of many fine chemicals can be produced in laboratory conditions using only a few production steps. The problem is that many of the most efficient reactions cannot easily be scaled up for large-volume production, as reaction conditions under larger-scale production are different from those at a laboratory scale. An exothermic reaction that works in the laboratory, for example, may not be usable today at a commercial scale because the relationship between the generation and dissipation of heat depends on the volume of the reaction vessel. A perfectly safe reaction in a laboratory test tube could lead to an explosion if scaled up.

As a result, only a small fraction of known reactions, around 20%, can be used at a commercial scale. This has a direct consequence for materials use and waste generation. As more direct production routes that are possible theoretically but are not possible in practice cannot be applied in today's production facilities, commercial production processes for many fine chemicals involve long, multi-stage sequences of reactions. In turn, for every additional reaction step there are buffering and other associated processes that use materials and produce wastes. The proliferation of reaction steps adds to inefficiency.

Solutions to the problem lie in three synergistic innovation lines. The first is to improve the basic toolkit of important conversion steps. The second is to develop solid-state catalysts and reaction media, so that each process step does not generate wastes from the auxiliary materials used. The third is to redesign commercial fine chemicals facilities and operations so that more of the known and efficient reactions can be used in commercial production. As the inefficiencies in commercial production arise not because more efficient reactions and production routes are not known but because these do not lend themselves to upscaling, the need is less for new knowledge about feasible reactions and more for new approaches to allow existing knowledge to be better applied. In turn, this implies that how commercial production is organised, including the architecture and logistical arrangements within fine chemicals production facilities, needs to be revisited from the sustainability perspective.

Today's arrangements are an outcome of 'borrowing' chemical engineering technology from the bulk chemicals sector. This is largely based around the concept of continuous production systems architecture. At the outset, when the fine chemicals industry was small, it may have been an appropriate economy to borrow engineering design and architecture from its big brother industry. This is no longer the case and is now actually detrimental. As the industry has grown in importance and its product range has diversified, a more customised set of chemical engineering arrangements based around batch production in quasi-laboratory-scale conditions would be more appropriate. In principle, improving materials efficiencies and avoiding the needless production of large quantities of waste depends on rethinking the whole design and operating arrangements within fine chemicals facilities. This opens a completely new avenue for innovation and R&D, which has potentially important spin-off benefits for the bulk chemicals sector too.

Box 9.2 **Improving efficiencies in precision chemistry**

species, whether these are grown in the Netherlands or elsewhere. It is expected that products from large-scale cultivation will enter trade and ultimately be available to Dutch industry.

In the field of fine chemistry, efforts are under way to rethink the production engineering and operating procedures at fine chemicals facilities in line with the principle that these should be customised to the needs of the industry and its scales of production rather than modelled on facilities and operations in the bulk sector. This would enable a wider set of reactions to be used and provide for a shortening of production routes. In addition, work is under way to enlarge the toolkit of reaction processes available to the fine chemicals sector and to develop new solid-state reaction media and catalysts. It is expected that these developments in the field of precision chemistry will lead in the long run to a much higher efficiency from chemical reactions and, therefore, to dematerialisation and a substantially lower environmental strain.

Chapter 10

Elaborating and evaluating opportunities: the case of organic materials

This chapter deals with organic materials and liquid fuels. Here the key question concerns the prospects for using renewable biomass as an alternative to fossil carbon feedstock when making organic materials, such as fibres, films and plastics, and when making liquid fuels. The question has two elements. The first is whether, in principle, enough biomass can be produced in sustainable ways to meet anticipated future needs for fuels and materials as well as for food. The second is to describe production routes by which the full range of major bulk synthetic organic chemicals and fuels can be made from biomass. The first of these questions was addressed in the strategic orientation phase of the STD chemistry project and it was found that, in principle, there is a sufficient biomass production potential to meet the anticipated future demand for industrial organic chemicals after the more pressing need to produce food has been met (Okkerse and van Bekkum 1996). Nonetheless, the biomass shortfall in respect to future fuel energy needs suggests that there is need to use biomass efficiently in all applications and to augment conventionally produced biomass with biomass produced in unconventional ways and with direct energy captured using photovoltaic technologies.

This chapter describes how the technological potential of a shift to using renewable resources in the organic chemicals sector was evaluated within the STD programme and how the implications of such a shift were drawn out for use in building a near-term R&D agenda. Special emphasis is placed on the need to provide detailed descriptions of feasible production routes between renewable biomass raw materials and sustainable products. Much of the discussion relates to the need to find ways of using the bulk of the available biomass to produce bulk commodity products. This is needed if serious inroads are to be made toward achieving sustainability in the organic chemicals sector. In order to make full use of all biomass, rather than only the small fraction that is used today, ways have to be found of valorising lignocellulosic materials and wet biomass. Potential solutions lie in the field of C_1 chemistry, i.e. chemistry based on molecules containing a single carbon atom. These solutions need to be developed alongside other routes to valorising biomass based on first extracting any high-value components and then breaking down remaining wastes to C_1 forms. The complement of this approach is the need to develop production routes for synthesising organic products from C_1 starting compounds. Other topics addressed in the chapter include the development of plant strains that can be grown in brackish soil as a way of increasing the quantity and quality of biomass available for chemicals production. The chapter draws on work by Professors Okkerse and van Bekkum within the context of the STD project on chemicals and industrial materials (Okkerse and van Bekkum 1996).

THE NATURE OF ORGANIC CHEMICALS PRODUCTION

Material products are end-points of production chains. The value-adding role played by the organic chemicals industry is to build successively more complicated structures of a desired type into chemical products through each successive process step. The basic building blocks for organic products are atoms of carbon. Building desired molecular structures from these and other elements requires energy. In today's organic chemicals industry, the carbon and energy are mostly derived from fossil carbon feedstock. However, in principle, both energy and carbon could equally be derived from renewable biomass feedstock. Indeed, there is a compelling analogy between the materials transformations achieved by the chemicals industry and those achieved in nature by organisms, especially by green plants during photosynthesis (Box 10.1). Through the metabolic processes of organisms, nature produces its own structures based on carbon as a building block for complex organic molecules. Some of the products, such as polysaccharides, disaccharides, triglycerides, polyterpenes and monoterpenes, are identical to those produced via industrial synthesis. Other molecular structures found in nature, such as cellulose, starch, sugars and monosaccharides, are within one or two biological or chemical process steps of final industrial products.

By implication, this gives three different routes for using biomass in the production of organic industrial materials and fuels. By the first route, constituents that are identical to the required final product can be extracted from plants and organisms and used directly as an alternative to production through chemical synthesis. By a second route, we can separate out those constituents that have a chemical structure close to the required final product. We can then use this

BIOMASS IS THE PRODUCT OF PHOTOSYNTHESIS BY WHICH GREEN PLANTS USE solar energy, water and carbon dioxide to produce carbohydrates. Light energy is used to split carbon dioxide obtained from the atmosphere into its chemical constituents. The carbon, together with water obtained from the soil, is constituted into carbohydrate molecules that form new plant tissues and into chemicals needed for the plant's life processes. The liberated oxygen is released as a waste product. The general form of the reaction is:

$$nCO_2 + nH_2O \rightarrow (CH_2O)n + nO_2$$

The theoretical maximum efficiency of photosynthesis is 6.6% (Barsham 1983). In practice, the achieved capture rates for light energy are substantially lower, 2.4%–3.2% for starch and sugar-rich C_4 plants (such as sugar cane, sorghum and miscanthus) and 1.7%–1.9% for C_3 plants (such as sugar beet and eucalyptus).

Plant tissues therefore represent a source of carbohydrate molecules that can be used as foods, materials and energy. Although the quantity of solar energy captured by plants and related organisms each year is small in comparison with the amount of solar energy received at the surface of the Earth, the quantity is large compared with the total annual anthropogenic energy use. This is an important consideration when evaluating the technical potential of biomass to replace fossil fuels as raw material sources for organic materials and fuels production. The future prospects for biomass will depend on the comparative economics of carbohydrates and hydrocarbons. The situation is already changing because advances in the materials and biological sciences are reducing the costs of manufacturing biomass-derived products while environmental regulations are beginning to increase the cost of hydrocarbon-based products (Morris and Ahmed 1992).

Box 10.1 **Photosynthesis and biomass**

backbone structure as a foundation to be 'built on' further in a form of semi-synthesis. The third route is a back-to-basics approach that involves breaking the biomass down into fundamental chemical building blocks and then rebuilding entirely new structures from these. The extent to which the materials are broken down can be controlled. In effect, then, this third route represents a continuum. At the very extreme, biomass may be broken down to C_1 forms, i.e. into molecules containing single carbon atoms. However, this may not always be necessary or desirable. For the production of fuels, for example, starting compounds of a form similar to petrol (C_5–C_{10}) may be more useful.

The important point is that these three routes are complementary: together they offer an approach for valorising all biomass. A rational use of biomass involves complete use, including the use of all different types of biomass material and the integral use of whole plants. High-value products and constituents need to be extracted from plants first for use as foods or industrial materials. Products, such as natural rubber, sucrose, lactose, agar and pectin, can be marketed directly. Constituents such as oils and starches can be extracted and used in one- or two-step conversions to produce products of semi-synthesis. The remaining waste material, together with entire plants that contain no high-value constituents and organic wastes from other sources (such as sewage sludge), should travel the third route and go for a primary treatment to produce starting compounds for chemical synthesis. Different primary treatment processes, yielding different starting compounds, are suitable for different kinds of waste. Especially, a distinction needs to be drawn between wet and dry wastes. Wet wastes are suitable for bio-crude production using hydro-thermal upgrading and for supercritical water gasification to produce hydrogen and methane. Biomass wastes containing less than 20% water (by weight) are suitable for flash pyrolysis. Dry wastes can be gasified to produce synthesis gas, a mixture of H_2 and CO, from which methanol can be produced. Methanol is an ideal basic chemical for use in the energy sector (for example, as a fuel for gas turbines) and in the chemical industry as a basis for C_1 chemistry.

In the context of this complementarity, we can see why the third 'back-to-basics' route is so important both for the economics of organic chemicals production from biomass and, also, for sustainability. A substantial amount of biomass waste is produced during the first and second routes. The yield of cellulose fibres from logs, for example, is 35%–40%. The yield of oil from oilseed rape is 40%. Typically more than half of the biomass does not proceed to the next process step. The wastes—lignin from pulpwood, bagasse from sugar cane or molasses from sugar beet—are available, in principle, to be valorised through the third back-to-basics route. The back-to-basics route also offers a way of making use of the greater part of biomass that does not contain any directly useful molecules and is not suitable for semi-synthesis. Lignocellulosic material that is not easily fermented falls into this category. In addition, some of the back-to-basic technologies offer a way of valorising wet biomass and organic wastes produced during food processing or at sewage works.

The significance of the back-to-basics route for sustainability becomes even clearer when analysed in the context of the scale of production of organic industrial materials and fuels. Almost 93% of the fossil energy and carbon used today is used to produce fuels or power. Only 7% is used to produce non-fuel industrial organic chemicals. In turn, only a few industrial organic chemicals are produced in high volumes. The top ten industrial organic chemicals account for more than 95% of all industrial organic chemicals production. By implication, if environmental stress is to be reduced measurably and if the organic chemicals sector is to make meaningful strides toward sustainability, biomass needs to replace mineral oil as the raw material used to produce liquid fuels and bulk industrial organic chemicals.

TODAY'S ORGANIC CHEMICALS INDUSTRY

Table 10.1 lists the major products of the organic chemical industry today and annual production volumes.[1] The top three bulk products are polyethylene, polypropene and polyvinylchloride. All of these are produced in quantities in excess of 20 million tonnes a year. They are produced from the lower-olefin monomers, ethane, propane and vinyl chloride. A fuller range of product categories, including speciality products is given in Figure 10.1, which also shows some basic production paths to final products from plant constituents, such as starches, sugar and oils. Tables 10.2 and 10.3 follow this line further. Table 10.2 summarises

Table 10.1 **The world's top ten organic industrial chemicals**

PRODUCT	WORLD PRODUCTION VOLUME IN 1994 (million tonnes)	MONOMER AND/OR CONVENTIONAL PRODUCTION ROUTE	WORLD MONOMER PRODUCTION IN 1994 (million tonnes)
Polythene	44.0	Ethane	70.2—used also for ethane oxide/glycol, ethylbenzene/styrene
Polypropene	20.0	Propane	39.5
Polyvinylchloride	24.0	Vinyl chloride	
Methanol	18.4	Natural gas	
Polyalkylene-terephthalate	13.1		
Polystyrene	12.6	Styrene	
Ethane oxide	9.6	Ethane	8.5 of the total 70.2 million tonnes ethane production (see above)
Butadiene	8.1		
Phenol	5.3		
Nylon	3.4	Caprolactan	
	2.0	Adipic acid	
	1.2	Hexamethylene-diamine	

1   Pulp and paper products are not included in the table, as these constitute a separate industry in their own right and one that is already based on biomass feedstock. There are 200 million tonnes of pulp and paper products produced annually compared with 100 million tonnes of products from the chemicals sector produced from fossil raw materials, excluding fuels.

PLANT MATERIALS ⟩ ⟩ ⟩ ⟩ ⟩ **PRODUCTS**

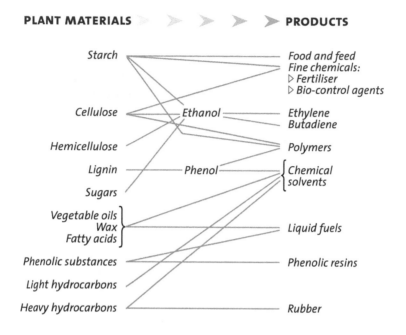

Figure 10.1 **Some first- and second-order production routes to bio-products**

Source: Moser 1998

Table 10.2 **Estimated output volumes and market shares of plant-derived products in the US market**

Sources: Moser 1998; USOTA 1992

PRODUCT	ALL PRODUCTION (10⁶ tonnes)	PLANT-DERIVED (10⁶ tonnes)	PLANT-DERIVED MARKET SHARE (%)	COST (US$/kg) FOSSIL	PLANT
Plastics	30.0	1.3	4.3	0.5	4.0
Pigments	15.0	1.4	9.0	2.0	5.8
Detergents	12.6	2.3	18.0	1.1	1.7
Wall paints	7.8	0.7	9.0	0.5	1.2
Adhesives	5.0	2.4	48.0	1.6	1.4
Dyes	4.5	0.8	15.0	12.0	21.0
Surfactants	3.5	1.8	50.0	0.5	0.5
Inks	3.5	0.6	16.0	2.0	2.5

the market share of plant-derived products across eight broad product categories, each with a production volume in excess of 3.5 million tonnes. Table 10.3 gives information about the main oil plants and the chemical products produced from them.

At first glance, these tables seem to suggest a broad biomass-based production potential. Nonetheless, the situation is not so straightforward as it might appear. Whereas, in principle, it is possible to produce final products from biomass, the number of cases where this actually happens today is limited mostly to speciality chemicals where few production steps are required. In the organic chemicals sector, the only competitive bio-products today are small-volume speciality products obtained directly from plants (first route) and a handful of products derived by semi-synthesis (second route), such as bio-glycerol and bio-alcohol or ethanol. The

Table 10.3 **Yields, prices and illustrative industrial uses for oil crops (actual and potential) in the US**

Sources: Moser 1998; USOTA 1992; Robbelen *et al.* 1991

OIL CROP	OIL YIELD (tonnes/ha)	OIL PRICE (US$/kg)	SOME MAIN USES
Bladderpod	3.9	–	Plastics, fatty acids, surfactants
Buffalo gourd	5.1	–	Epoxy fatty acids, resins, paints, adhesives
Castor	2.3	0.80	Dyes, paints, varnishes, bio-pesticides
Coconut	8.0	–0.46	Polymer resins, cosmetics, pharmaceuticals
Corn/maize	7.0	0.62	Fermentation products
Crambe	3.0	1.55	Paints, industrial nylon, lubricants, plastics
Cuphae	4.0	–	Surfactants, lubricants, glycerine
Euphorbia	4.5	–	Surfactants, lubricants, paints, cosmetics
Honesty	4.0	1.53	Plastics, foam suppressers, lubricants
Jojoba	8.3	9.60	Cosmetics, pharmaceuticals, inks, plastics
Lesquerella	1.8	–	Paints, lubricants, hydraulic fluids, cosmetics
Linseed	2.1	0.50	Drying oils, paints, varnishes, inks
Meadowfoam	3.2	–	Cosmetics, liquid wax, lubricants, rubber
Oil palm	12.5	0.34	Fermentation products, soap, wax
Rapeseed	4.0	1.30	Plastics, foam suppressers, lubricants
Safflower	2.8	0.80	Paints, varnishes, fatty acids, adhesives
Soybean	1.9	0.40	Inks, paint solvents, plasticisers, resins
Stokes Astor	3.9	–	Plastic resins, plasticisers, paints
Sunflower	4.3	0.32	plasticisers, fuel additives, agrochemicals
Vermonia	1.7	1.60	Plastics, alkyd paints, epoxy fatty acids

biggest, and best-known, industrial use of biomass is for making alcohol (ethanol) by fermentation of sugar using yeast. In Brazil, ethanol is fermented directly from cane sugar for use as an automotive fuel. In the US, ethanol fermented from glucose obtained from cornstarch by enzymatic conversion is used to make the fuel additive ethyl-*t*-butyl-ether (ETBE).[2] However, even though bio-ethanol production is nationally significant in these two countries, it is not significant globally.

More generally, bio-products are confined today to relatively small applications areas and to the specific niche of high-value, low-volume product markets. Although the role of bio-products is significant and growing in this market segment, especially in the US (Robbelen *et al.* 1991),[3] the speciality chemicals segment is too small for substitution within it to make a major contribution to sustainability. However much substitution occurs in speciality markets, this will not contribute substantially to sustainability as substitution here will not affect the use of fossil raw materials in the mass applications areas of bulk organic industrial chemicals and fuels production. For any real inroad to be made on sustainability, ways must be found for making bulk fuels and the bulk commodity chemicals listed in Table 10.1 from the lignocellulosic materials that constitute the bulk of the available biomass, and of doing so at low cost. This implies competing in large-volume, low-margin, commodity chemicals markets where prices are typically US$1–3 per kilogram of product.

Effectively, then, for the organic chemicals sector as a whole to use renewable resources, economically competitive production routes need to be found for making commodity products from lignocellulosic raw materials. This explains the importance of the third, back-to-basics, route. Unlike other routes, it offers an opportunity to use the bulk of available biomass for the synthesis of bulk products, rather than using just specific components such as starch or sugar. The technological challenge that this implies reduces to two basic elements. There is need at the front end of the materials chain for primary conversion technologies for breaking down the complex and unwanted structures in lignocellulosic plant materials

---

2   Another important fuel additive is methyl-tert-butyl-ether (MTBE). MTBE is the fastest-growing organic chemical product of the last ten years. Demand for it has grown because of the need to substitute tetra-ethyl lead as an octane-enhancer.

3   Robbelen *et al.* note that, in the US, over the past decade, there have been major shifts in market share in almost all product categories toward bio-products. The total market for bio-products was anticipated to grow to 10 million tonnes by 1996 from a base of 5 million tonnes in 1990. The major factor has been a strong and continuing trend of price reduction, which has lessened (but not yet removed) the price premium on bio-products in many product categories. The market shares of bio-inks, bio-detergents and bio-plastic in particular have grown strongly. The market for soy-based ink has been boosted by the trend toward fibre recycling. By the mid-1990s, soy-based ink accounted for more than 75% of the newsprint market and 50% of the magazine market. Many of the successful bio-products are produced from vegetable oils, squeezed from oil-rich plants. More than 20 different oilseed crops are grown in the US, with soybean dominating production. About 1 million tonnes of vegetable oil are now used as an industrial feedstock.

and releasing the constituent chemical building blocks. The complementary need is to describe new synthesis routes for making today's products from these basic starting compounds.

PRIMARY CONVERSION TECHNOLOGIES

Primary conversion of biomass can be achieved using several known technologies depending on the water content of the substrate (Fig. 10.2). Most of these fall into one of two groups depending on whether the primary conversion is based on a thermo-chemical or a biological treatment.[4] The main thermo-chemical technologies are combustion, gasification and thermal decomposition. The main biological technologies are fermentation and anaerobic digestion. Figure 10.2 shows the technologies and their major products. Most have severe disadvantages when the primary objective is to use biomass to produce materials rather than energy. Either the major products are heat and combustible gases rather than starting compounds or the technology is suitable for converting only selected types of biomass.

▷ **Direct combustion.** Direct combustion is the simplest conversion technology and is still a preferred route for converting mixed or variable feedstock into hot gases as a source of process heat, to raise steam or to generate electricity. It is also used, with or without heat recovery, to incinerate some problematic wastes. Modern, large-scale biomass combustion involves automated, fluidised bed technology originally developed for high-sulphur-coal burning facilities. The technology is quite clean since, by adding limestone to the fluidised bed, the SO_2 released during combustion is bound to $CaSO_4$. As an additional advantage, the NO_x emissions are limited due to the relatively low combustion temperature (c. 850°C). Nonetheless, combustion is inherently limited in that heat and power are the only products.

▷ **Gasification.** Gasification is a high-temperature process that is steered by controlling the temperature and the supply of the gasification agent. It is used to provide a combustible gas for power generation. The carbon content of the biomass is gasified by partial oxidation to CO at around 900°C in the presence of a limited supply of air or oxygen. In the basic process, air is the gasification agent. The product stream is a fuel gas mixture of H_2 (8%–18%), CO (16%–24%), CH_4 (2%–6%), CO_2 (9%–15%) and

---

4   A technology that does not fall easily into these categories is the extraction of vegetable oils from crops such as oilseed rape, which can be refined to produce bio-diesel. There is also a hybrid category based on bio-chemical conversion technology.

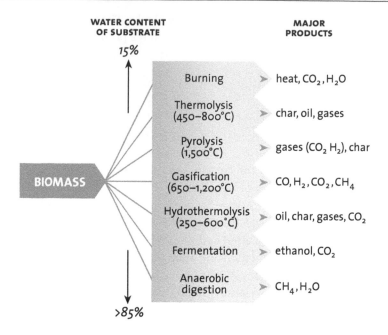

Figure 10.2 **Biomass conversion technologies**

N_2 (44%–52%) with a calorific value of the order of 6 MJ/Nm³. Contamination by tar is a serious problem. The fuel gas mixture is mostly burned directly to produce heat, but can be cleaned and used to generate electricity, although there are still technical difficulties in ensuring that the produced gas is free from dust and tar contaminants (TERES 1994).

▷ **Thermal decomposition.** Thermal decomposition of biomass involves heating the biomass in the absence of air at temperatures from 300°C. The product distribution depends on the applied temperature and the particle size. A low temperature and large particle size leads to charcoal as the main product. High temperatures and small particles are favourable for the production of gases. In the intermediate range, the production of condensible vapours is possible (tars of bio-oil).

▷ **Fermentation.** Fermentation or bioconversion is a long-standing technology in which sugars are converted to alcohol (bio-ethanol) by microbial action. However, this technology is restricted today to a small subset of available biomass that is rich in starches or sugar.

More promising than the established technologies are three new primary conversion technologies—advanced gasification, hydro-thermal upgrading and flash pyrolysis. These are all at an early stage of development and need to be both improved and upscaled. Nonetheless, in principle, they offer the possibility of converting bulk biomass and organic waste into versatile liquid or gaseous starting materials from which whole families of final products could be produced.

▷ **Advanced gasification.** Advanced gasification processes use oxygen-enriched air or pure oxygen to obtain a higher degree of control over the output profile than in conventional gasification. Steam can be added as a reagent or to control the temperature. This produces a much higher-quality gas with a calorific value of 10–14 MJ/Nm^3 in the case of oxygen-enrichment without steam or 14–20 MJ/Nm^3 in the case of oxygen-enrichment with steam added. The product stream—synthesis gas—is purer from advanced gasification and can be used directly for methanol or ammonia synthesis (TERES 1994).

▷ **Flash pyrolysis.** Flash pyrolysis offers a way of producing bio-oil from biomass. The basic principle of flash pyrolysis is to apply a temperature treatment sufficient to de-polymerise cellulose, but insufficient to cause a total breakdown to the C_1 level. The objective is rather to retain as much integrity as possible in the de-polymerised fragments, while still achieving de-polymerisation. This is achieved by a very short heat treatment (1–5 seconds at 500°C). The main product is bio-oil containing C_5–C_{10} fragments, such as phenols, levoglucosan and hydroxyacetaldehyde. The bio-oil contains some water and also significant quantities of acetic and formic acid. Char is another by-product. There is also a gaseous product stream (Box 10.2).

▷ **Hydro-thermal upgrading (HTU).** HTU involves high-temperature and high-pressure treatment of biomass and water to produce bio-crude, an equivalent to crude oil (Box 10.3). The technology has been under development by Shell at a laboratory scale in Amsterdam since 1988. It involves pre-treating a slurry of biomass and water in a digester at 200°C and 30 bar before passing it through a series of autoclave reactors in which the temperature and pressure conditions are raised to 330°C and 200 bar. The contained oxygen is driven off under these conditions, which are similar to those under which crude oil would form naturally over geological time. The yield is bio-crude with a low content of oxygen.

RECONSTRUCTING
FINAL PRODUCTS

Is it possible to identify production routes for all the major organic industrial materials and fuels using destructured starting materials obtained from biomass?

THE BASIC PRINCIPLE IN FLASH PYROLYSIS IS THAT, BY RAPID HEATING OF lignocellulosic feedstock in the absence of oxygen, it is possible to achieve a very fast but controlled chemical degradation of the polymer structure. Effectively, the feedstock can be de-polymerised without returning the contained carbon to C_1 forms. Rather, the polymer chains are broken while preserving the integrity of the individual links making up the chain (Fig. 10.3). This involves rapid but short heating of feedstock particles in the absence of air. To achieve rapid heating, the cellulosic feedstock must first be dried (<10% moisture content) and broken into small particles (< 3 mm) before entering the reactor. To limit the extent to which the materials are de-structured, the residence time of the feedstock and products in the reactor need to be short.

The reaction by which the feedstock is broken down consists of two separate process stages. The first is wood pyrolysis or de-polymerisation, in which the wood is reduced to liquid, gaseous and solid fractions but where each fraction consists of materials that maintain their original structural integrity. The second step, which would occur if the materials were allowed to stay within the reactor, involves the cracking or decomposition of the liquid fraction. The purpose of flash pyrolysis is to achieve the first step (de-polymerisation), but to avoid the second step. Since in both steps the extent of the reaction is a function of the temperature and residence time, there is need for a strict control of both to achieve de-polymerisation without decomposition.

Yields of the different fractions (bio-oil, gas and char) are also temperature-dependent (Fig. 10.4). The char yield decreases and the gas yield increases as the temperature increases (Wagenaar et al. 1993). De-polymerisation can occur over a wide range of temperatures, but the maximum yield of bio-crude (79%) has been found to occur at around 500°C and 1 atmosphere. A typical yield per 1,000 kg of wood (10% moisture content) includes 680 kg of bio-oil, 230 kg of gas and 90 kg of char. The produced bio-oil has a volumetric energy density of 21 GJ/m^3, a density of 1220 kg/m^3 and a viscosity at 50°C of 13 mm^2/s. The water content of the oil is 20% (by weight), the ash content 0.02% (by weight). The bio-oil is pH3 (Gros 1995).

There are several currently available technologies, including fluid bed (US/Spain), circulating fluid bed (Italy) and vacuum pyrolysis (Canada). A rotating cone reactor vessel has been used in Dutch trials (Fig. 10.5). The cone provides a large surface area for heating while its rotation forces both feedstock and reaction products quickly through. There is no carrier gas. There is a high heat transfer and a high rate of solids throughput. The residence time of solids and gases in the reactor can be controlled from 0.1 to 5.0 seconds. The main advantages are that this is a simple process, investment costs are low and there is low contamination of the final product. The process is simple and cheap, especially in comparison with the HTU high-pressure process; a recent report estimates US$17 per barrel for the production costs of bio-oil at a scale of 720 tonnes of feedstock per day and presumed feedstock costs of US$50 per tonne (Prins and Van Swaaij 1998). Because bio-oil is the major product, it can be used in subsequent stages to produce both fuels and materials. The process is proven at the pilot plant scale, but there is need for further R&D, scale-up, demonstration of the technology and business appraisal. R&D is needed, *inter alia*, into pressured pyrolysis, *in situ* hot filtering, coupling of the reactor to small turbines and chemical recovery/separation technologies. It is envisaged that scaling-up could be achieved by stacking cone reactors (Prins and Van Swaaij 1999).

Box 10.2 **Flash pyrolysis**

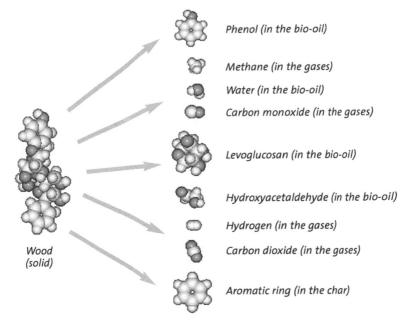

Phenol (in the bio-oil)

Methane (in the gases)

Water (in the bio-oil)

Carbon monoxide (in the gases)

Levoglucosan (in the bio-oil)

Hydroxyacetaldehyde (in the bio-oil)

Hydrogen (in the gases)

Carbon dioxide (in the gases)

Wood
(solid)

Aromatic ring (in the char)

ORIGINAL WOOD STRUCTURE
(at 20°C)

SOME PRODUCED FRAGMENTS
(after pyrolysis at 500°C)

Figure 10.3 **Flash pyrolysis principle and products**

Source: University of Twente

Figure 10.4 **Distribution of primary products**

Source: Wagenaar 1994

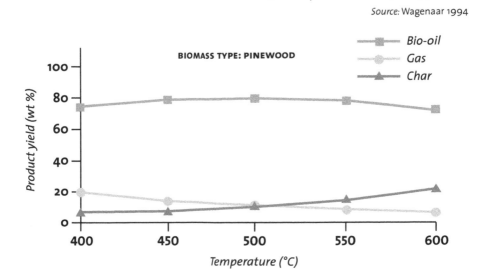

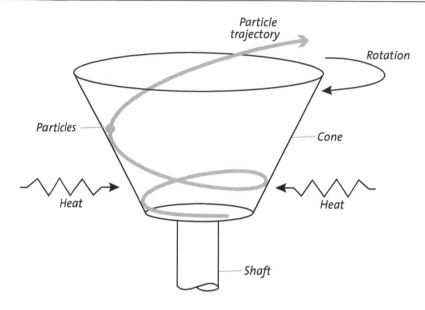

Figure 10.5 **Principles of reactor operation**

Source: University of Twente

Clearly, in respect to two of the most promising primary conversion technologies, flash pyrolysis and hydro-thermal upgrading, the bio-oil and bio-crude output from primary conversion is similar to today's fossil raw materials. Although there are technological obstacles still to be overcome, these technologies could potentially offer the possibility to continue along 'business-as-usual' lines in today's petrochemicals sector, with the exception that the basic raw material is renewable rather than of fossil origin.

The output from advanced gasification, the other promising back-to-basics option, is a mixture of hydrogen and carbon monoxide gases known as synthesis gas or 'syngas'. Syngas is a starting mixture for producing hydrogen-rich fuels.[5]

5   The possibilities for producing fuels from syngas are described in greater detail in Chapter 11. In some applications, synthesis gas needs first to be enriched with hydrogen. The question arises, therefore, of how hydrogen might be obtained in the future on a sustainable basis. There are several possible routes. One of the more important from the perspective of technological innovation is by electrolysis of water using renewable electricity. Photovoltaic cells constitute one of the most promising emerging technologies here, directly converting sunlight into electricity. Considerable progress has been made already on improving the efficiency and lowering the cost of electricity production using photovoltaic cells. Many different variants of cell technologies and production routes for these are now known. As a consequence of this variety, and because photovoltaic cells are a nascent technology, the prospects for further substantial cost and performance breakthroughs are excellent. Although much work has still to be done in this—arguably the key—area of sustainable technology development, it is one where we can expect breakthroughs to be made.

HYDRO-THERMAL UPGRADING (HTU) POTENTIALLY OFFERS A NUMBER OF advantages. the process has a high yield and a high energetic efficiency. Many different kinds of organic matter can be used as feedstock, including organic wastes such as sewage sludge. Since the process requires a mixture of water and biomass (in the ratio 3:1), there is no need for the biomass to be dry. (In contrast, dry biomass is needed for advanced gasification.) All of these make it a potentially interesting process, although the costs of the process equipment are likely to be high because of the high pressures and long conversion times (10–15 minutes) involved. Yields of 40% have been achieved in laboratory-scale tests using wood, agricultural waste, household waste and sewage sludge as feedstock. Using sewage sludge opens the possibility of a zero or even negative raw material cost as the process operator could receive a payment equal to the costs of conventional disposal, which are rising.

Economic projections for a commercial plant costing US$105 million and having a capacity to process 1,440 tonnes of feedstock and yield 625 tonnes of bio-crude daily suggest a bio-crude production cost of US$20–40 per barrel (Goudriaan cited in Okkerse and van Bekkum 1996). This range reflects different assumptions about feedstock costs, which, in turn depend heavily on the logistics arrangements surrounding the delivery of feedstock to the facility. If the feedstock has zero cost, the bio-crude production cost could be as low as US$20 per barrel. However, this ignores the potential 'negative value' of sewage sludge, which could lower the production cost to a level below US$20. While not yet competitive with conventional crude, unless based on a negative value feedstock, bio-crude could quickly become competitive. If successfully scaled up so that a 40% yield is maintained or exceeded in commercial-scale production, bio-crude could be competitive at a price for conventional crude for which there are historical precedents. If a premium equivalent to the cost of sewage sludge disposal were paid to the bio-crude producer, the process would be competitive even at the long-term average oil price, which is between US$15 and US$20 per barrel. On an energy-equivalence basis, bio-crude is also cheaper to produce than either bio-ethanol by fermentation of starches and sugars (US$100 per barrel) or methanol from gasification of biomass (US$65 per barrel), though not as cheap as bio-oil produced by flash pyrolysis.

Box 10.3 **Hydro-thermal upgrading**

Equally, it can be used to produce methanol. The first requirement is therefore to demonstrate production routes by which bulk commodity products can be reached from methanol and other plant-derived starting compounds. Following this, there is a need to identify what changes to production processes will be required and what key technological bottlenecks and opportunities exist in respect to these as the basis for deciding near-term R&D priorities. The STD programme reviewed all of the top ten commodity organic chemicals to establish feasible production routes for producing these from biomass, focusing especially on synthesis routes based on C_I starting compounds.

The highest production volume synthetic organic chemical is ethane (C_2H_6), the monomer used for the production of polyethylene (C_2H_4). Ethane is one of the lower olefins, a group that also includes butane (C_4H_{10}) and propane (C_3H_8). In principle, there are at least two routes to the lower olefins from biomass other than

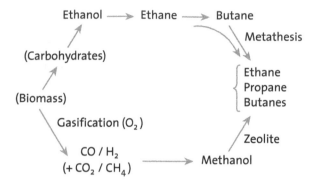

Figure 10.8 **Bio-based routes to lower olefins**

by recourse to bio-crude (Fig. 10.8). One is based on fermentation of carbohydrates to produce bio-ethanol (C_2H_6O). As already indicated, bio-ethanol is produced today as a fuel and/or fuel additive in Brazil and the USA. In India, bio-ethanol is also used to produce ethane for making polythene and for making ethylbenzene ($C_6H_5CH_2CH_3$) and 1,4 diethyl-benzene. Because fermentation processes work well only on starch and sugar, this route is limited unless fermentation technologies are developed that enable cellulose and hemicellulose to be used as a feedstock. However, this may be neither necessary nor worthwhile, as there is a second route to the lower olefins via syngas and methanol. This second route involves the dehydration of methanol using zeolite-based catalysts[6] (Box 10.4).

The same two production routes, via ethanol produced by fermentation or via methanol produced by biomass gasification, apply also to the production of propane, the monomer for polypropene.[7] However, here, the advantages of the latter route are even more pronounced because it involves fewer steps and is therefore more efficient. To produce propane from ethanol is a multi-step process involving, first, the production of ethane and then a metathesis step to give propane. Alternatively, propane can be produced directly in one step from methanol by using a zeolite catalyst. The methanol route is therefore particularly attractive for propane production because it is shorter and more efficient than fermentation and production can be based on any biomass feedstock.[8]

6 A US firm, UOP, is developing the technology.
7 There is also a third possibility that involves the fermentation of carbohydrates to a combination of acetone and butanol. However, this third possibility requires either that the two products are used together or that they are separated before use, which is difficult today. Membrane technology may be developed to facilitate separation. Nonetheless, as with fermentation processes generally, this process is applicable only to starch- and sugar-rich feedstock.
8 Within this discussion it is important to consider whether substitutes for polyolefin films are likely to be developed. The most obvious alternative would be cellulose film. However, cellulose is not water-resistant. There is the possibility of coating cellulose film with

ZEOLITES ARE SPECIALISED TETRAHEDRAL STRUCTURES OF SILICON AND OXYGEN that may also contain aluminium. Silisium occupies the central position within the tetrahedron. Oxygen atoms occupy the outer sites. Extra tetrahedra can be added to form a zeolite lattice. The outer sites, occupied by oxygen, are shared in the formation of these additional tetrahedra, but each new tetrahedron creates a new central site. This can be occupied by silisium (4+) or aluminium (3+). If aluminium is introduced, additional cations are needed to render the structure neutral. These can be provided by, for example, sodium (1+), potassium (1+) or magnesium (2+). One of the essential characteristics of zeolites is that they offer spaces within their lattice structure where reactions and separations can take place. Moreover, zeolites can be synthesised specifically from the perspective of controlling the size of the spaces. Another key characteristic is that atoms of a desired chemical, such as a platinum catalyst, can be placed within the lattice structure to provide specific catalytic properties. This means that specific zeolites can be designed to act as chemical enzymes. Catalysis is achieved by designing a zeolite structure that interfaces well with, and which therefore weakens the internal bonds within, the structure of the material to be converted.

Box 10.4 **Zeolite catalysts**

Unlike the market for other polyolefin products, the market for polyvinyl-chloride, $(CH_2CHCl)_n$, is not growing. New technologies have begun to displace polyvinylchloride (PVC) in some of its main application areas, such as for making door and window frames. Nonetheless, the potential for using inert PVC for high-value applications—for example, for making pipes for delivering drinking water—is good and there is likely to be a continuing, strong demand for PVC in such applications (see also Chapter 7). PVC is made from vinyl-chloride monomer. In small-scale production, acetylene (CHCH) is used as a starting material. Larger-scale production uses a three-process-integrated production sequence based on ethane:

$$H_2C = CH_2 + Cl_2 \quad \rightarrow \quad ClH_2CCH_2Cl$$

$$ClH_2CCH_2Cl \quad \rightarrow \quad ClHC = CH_2 + HCl$$

$$H_2C = CH_2 + HCl + O_2 \quad \rightarrow \quad ClH_2CCH_2Cl$$

As already described, for renewable-based production, ethane can be derived from bio-ethanol, bio-methanol or bio-oil.

Methanol is itself a top-ten synthetic organic chemical. Today, production is mostly from natural gas (CH_4) or naphtha $(C_{10}H_8)$. In the case of CH_4 as a feedstock, H_2 is oversupplied and there is need for additional CO_2. In the case of naphtha, the ratio of H_2 to CO is directly suitable for methanol production. Methanol is used as

a water-resistant material, such as a bio-derived polyolefin. This would have the effect of reducing the overall market for polyolefin films. However, very thin coatings would probably depend on developments in nanotechnology. It is likely, also, that coated films would be expensive to produce. For the foreseeable future, therefore, polyolefin films are unlikely to be displaced on the market.

a fuel directly and also as a fuel additive. Almost 40% of output is used to make methyl-tert-butyl-ether (MTBE), which is used as an alternative to tetra-ethyl lead as an octane-enhancer in fuels. In a sustainable economy, both the method of producing methanol and its uses would change. Production would be from syngas obtained by biomass gasification. In addition to direct use as a fuel and fuel additive, methanol could play an important role in a future sustainable economy as a hydrogen-rich, liquid energy carrier useful for on-demand production of hydrogen for fuel cells by *in situ* catalytic reformation (see Chapter 11). In addition, methanol is an important starting compound for the production of a wide range of materials, such as acetic acid (CH_3OOH) and formaldehydes ($H_2C=O$). In respect to this versatility (Fig. 10.9), many analysts consider methanol to be a key chemical in the transition to sustainability.

Synthetic fibres based on polyalkylene-terephthalate constitute another major commodity group and one that is fast growing. Polyalkylene-terephthalate production is of the order of 13.1 million tonnes annually. This group is increasingly dominated by polyester, which accounts for about 60% of total production. Other important products are polyacrylic and polyamide fibres. Today, these products are derived from terephthalic acid, $C_6H_4(COOH)_2$, produced from *p*-xylene (C_8H_{10}). Alternative bio-based routes to terephthalic acid include production from the monoterpene, limonene, which is obtainable inexpensively as a by-product of

Figure 10.9 **Methane as a key chemical**

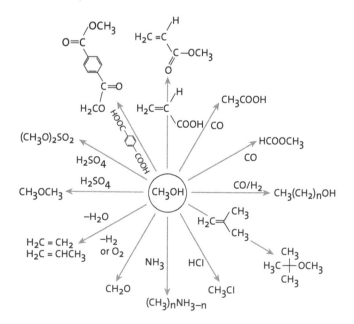

▷ **FROM INEXPENSIVE MONOTERPENES**

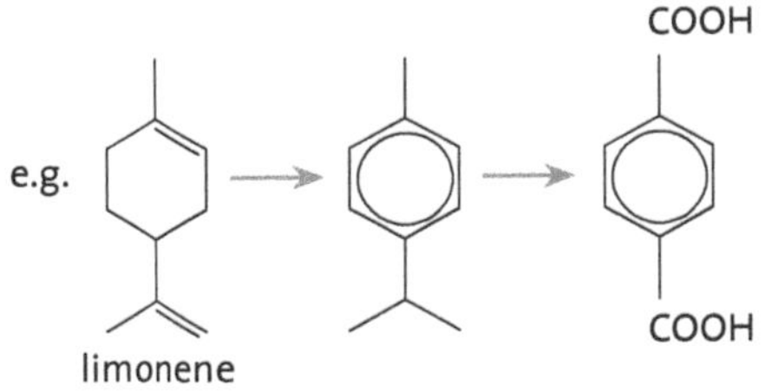

e.g.

limonene

▷ **FROM FRUCTOSE VIA HYDROXYMETHYLFURFURAL**

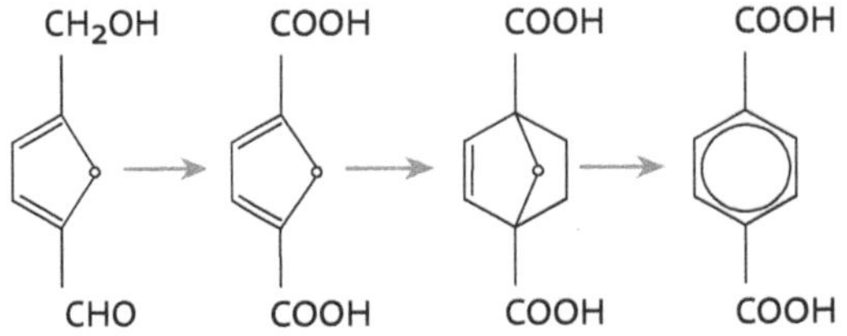

▷ **HTU AND PRESENT-DAY OIL REFINERY**
AND PETROCHEMICAL PROCESSES

Figure 10.10 **Bio-based routes to terephthalic acid**

citrus fruit and pulp/paper production,[9] and production from fructose via hydro-methylfurfural (Fig. 10.10). As with most other products, polyalkylene-terephthalate can also be produced from bio-oil using conventional technology. In addition to these possibilities, a close (and superior) substitute to polyester, polytrimethylene, can be produced from trimethylene-glycol, which can be obtained directly by fermentation of carbohydrates using genetically modified yeast.

Styrene ($C_6H_5CH=CH_2$), butadiene ($CH_2CHCHCH_2$) and phenol (C_6H_6O) are more difficult to make from biomass. Styrene is used directly to make polystyrene. After polystyrene, the biggest use of styrene is for co-polymerisation with butadiene to styrene-butadiene rubber (SBR). The annual production of SBR, poly-butadiene and poly-isoprene is around 11 million tonnes. It should be borne in mind that more than 5 million tonnes of poly-isoprene are derived directly from nature as natural rubber, an amount equal to one-third of total (natural and synthetic) rubber production. On a renewable basis, styrene can additionally be produced from butadiene by dehydrogenation over a copper catalyst. In turn, butadiene can be obtained in a multi-step process from bio-ethanol or in a one-step process from

9 Limonene is found in the peel of citrus fruits and also in pine trees.

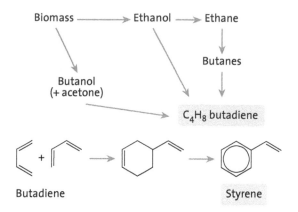

Figure 10.11 **Biomass-based route to styrene**

bio-butanol and acetone (CH_3COCH_3) using processes that have been described already (Fig. 10.11). A better option is to use the already-described route to produce butane from methanol using zeolite catalysts, followed by dehydrogenation to deliver butadiene.

Today, most phenol is either produced together with acetone from isopropyl-benzene (C_9H_{12}) or by catalysis from benzoaldehyde (C_7H_6O) produced via benzoic acid (C_6H_5COOH) from toluene ($C_6H_5CH_3$). Ideally, a short route to phenol from biomass is needed, but, because there is a relatively low proportion of phenol-like structures within natural biomass, phenol has to be synthesised in a series of process steps. Most production routes are, therefore, rather tortuous and have high costs. One such route is via butadiene and propane (Fig. 10.12). A more direct route is to produce phenol from fructose. Probably the best option, however, is from bio-oil produced using hydro-thermal upgrading or by flash pyrolysis. In a renewable economy, phenol and phenol-derived products are likely to be relatively more expensive to produce than other renewable products because of the low levels of naturally occurring phenol structures within biomass. This suggests that there is a need to search for substitutes for products of the classic phenol/formaldehyde system.

This has implications for the production of products within the nylon group of synthetic fibres. The present production process for nylons in the C_6 system uses the monomer caprolactam for nylon-6 and adipic acid and 1,6 diamino-hexane for nylon-6,6. In a renewable economy, caprolactam, $NH(CH_2)_5CO$, produced either from phenol or from cyclohexane, will come into a high price category. The problem is less for the nylon-6,6 monomers, which can be produced from buta-diene via ethanol or from butane (Weissermel and Arpe 1993). However, raw mate-rials for producing other products in the nylon sector could be drawn directly from plants and used for semi-synthesis. A chief component of castor oil, richino acid,

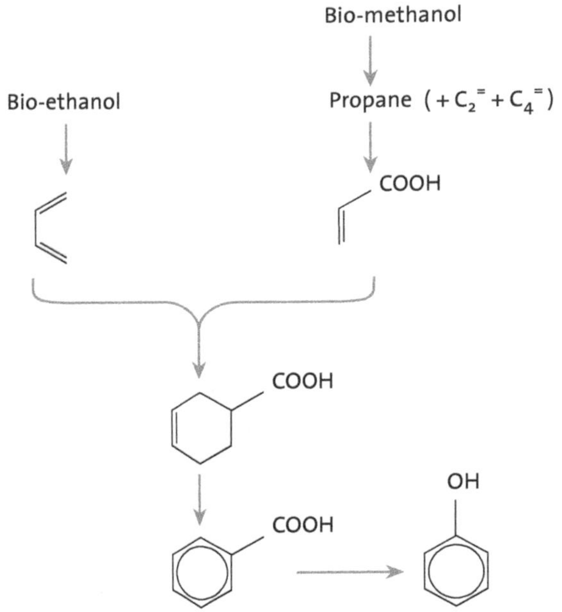

Figure 10.12 **Biomass-based route to phenol**

can be used to provide the base materials for nylon-11 while the monomer for nylon-10 can be produced in a single-step process from another naturally occurring plant material, oleic acid ($C_{18}H_{34}O_2$). Oleic acid (like bio-methanol) can be developed into a key chemical because several tri-glyceride-producing plants can be modified to produce it in high quantities. This suggests that R&D might usefully be focused on increasing the yield and extraction rates of such naturally occurring plant constituents as sources of raw materials for the chemistry sector.

AUGMENTING BIO-CHEMICALS PRODUCTION USING NEW BIOMASS

One important possibility in this regard is to extend cultivation of biomass to saline soils and to grow plants on such soils that are specifically developed to produce high yields of useful chemicals (Rozema *et al.* 1996). The context for this is set by the fact that the area of saline soil is increasing, especially because of poor or inappropriate irrigation practices. Around 30% of the world's agricultural soils are now classed as brackish, i.e. they have a salt concentration of 50–150 mM NaCl-equivalent. The problem is especially severe in low-lying irrigated areas near rivers and coasts because of the slow speed of underground transport of irrigation water away from the soil. When the movement of water is slow, there is the risk that water will evaporate from the soil surface causing an accumulation of salt residues. The risk is higher if insufficient irrigation water is used and if the irrigation water has a high content of dissolved salt. The accumulation of salt has an effect on the osmotic pressure around plant roots.

The mechanisms of impact of salty soils on plants are complex. There can be both primary and secondary stress impacts linked to the inability of plants to obtain water or nutrients from the soil or from the direct and indirect effects of salt accumulation within plant tissues. Figure 10.13 shows the influence of increasing salinity on the growth and/or yield of mangrove, sugar beet, grass, wheat and beans. The classification 'brackish' is represented in Figure 10.13 as the region bordered by the two thick shaded lines. At very best, the yields of many conventional crops such as cereals and legumes are severely reduced. As the concentration of salt in the soil increases, the yield of each plant decreases relative to the yield that would have been obtained under salt-free conditions. Compared with salt-free conditions, the yield of each plant at a salt concentration of 150 mM NaCl-equivalent is 90% for sugar beet, 80% for grass and less than 30% for wheat. At this salt level, beans cannot be grown at all.

Figure 10.13 **Salt yield relationship**

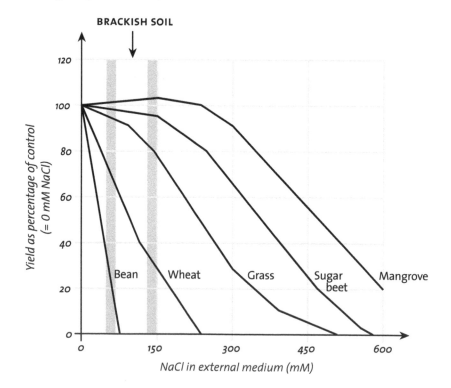

Of note in Figure 10.13 is that sugar beet, like mangrove, is relatively unaffected by salt in soils. This is because sugar beet (*Beta vulgaris*) belongs to the halophyte group of salt-tolerant plants. As a halophyte, an efficient converter of solar energy and a producer of molecular structures useful for chemical semi-synthesis, sugar beet is potentially a very important crop for use on saline soils. Its potential raises important questions. Is sugar beet the only commercial plant in the halophyte group? How many other halophytes are there? Are there other potentially useful natural halophytes? In respect to the first of these questions, Table 10.4 classifies and enumerates plant genera that contain a lot of halophytes (Flowers *et al.* 1986). It shows that the halophyte gene pool is quite rich, encompassing several families of plant, each with a good representation of halophytic plant varieties. In respect to the second question, there are both wild and cultivated halophytic plants that naturally yield useful materials. Garden marigold (*Calendula officianalis*) is rich in unsaturated hydrocarbons potentially useful as paint solvents. The African sea kale (*Cranberra Abyssinia*) contains acids potentially useful in making nylon, paints and coatings. Honesty (*Lanuria annua*) yields acids that could be used to make detergents and anti-foaming additives for washing powder. A form of sea-spurge (*Euphobia lacharska*) contains epoxy acid that could be used to stabilise PVC.

Brought under large-scale commercial production, plants such as these could yield substitutes for traditional raw materials of the chemicals industry and also

Table 10.4 **Halophytic plant families**

Source: Flowers *et al.* 1986

FAMILY	NUMBER OF HALOPHYTIC GENERA	NUMBER OF HALOPHYTIC VARIETIES	NUMBER OF GENERA WITHIN THE FAMILY	PERCENTAGE OF HALOPHYTIC GENERA IN THE FAMILY
Poaceae	45	109	650	7
Chenopodiaceae	44	312	100	44
Asteraceae	34	53	1100	3
Rizoaceae	21	48	143	15
Papilionaceae	19	35	700	3
Apiaceae	19	31	300	6
Euphorbiaceae	15	33	300	5
Brassicaceae	15	30	380	4
Cyperaceae	13	83	90	14
Arecaceae	13	22	212	6
Scrophulariaceae	13	21	220	6
Caryophyllaceae	9	16	80	11

do away with the need for some steps or even complete chain processes of traditional chemical synthesis. To make the connection to the chemistry sector, there is need to improve the salt-resistance of these natural halophytes and also the yields of their useful components. There is also the possibility of introducing elements of the DNA of natural wild halophytes into cultural varieties to produce salt-tolerant plants that directly yield high-level organic chemical products that do not require further chemical synthesis. Any such development and improvement of halophytic crops to supply raw materials and products for the chemicals sector needs also to be supported by related research to improve techniques for farming brackish soil. Because the use of chemical fertiliser, pesticide and herbicide is counter-indicated in arid and salty conditions, halophyte production should be based on bio-fertilisers, bio-pesticides and bio-herbicides. For this reason, there are strong synergies between research on plant genetics, bio-agriculture and the chemicals sector.

CONCLUSIONS

As argued in the previous chapter, to become more sustainable means reducing the overall annual draw on non-renewable resources and switching to renewable resources whenever appropriate. This means shifting from using fossil feedstock to using renewable sources of materials and energy. The changes needed in the technologies of production processes are therefore at the front-end of the materials chains in respect to the sources of raw materials, primary conversion processes and primary synthesis. This is in stark contrast with today's pattern of R&D expenditure, which is focused on those parts of the materials chain where most value is added. The majority of R&D expenditure today is devoted to process and product improvement in the middle and at the tail ends of the chain. Relatively little attention is given to the front end of the chain. The implicit logic that underlies this rests on the assumption that the raw material source is a factor neither in continuity of production nor in future competitiveness among producers. This assumption is unrealistic. This being the case, it is essential that the bulk organic chemicals and fuels sector initiates timely R&D to develop replacements for fossil sources of materials and energy.

Restructuring is not new to the chemicals sector. There have been several earlier transformations that have arisen from shifts in the basic feedstock. When coal was the main raw material source, acetylene and aromatics were the main starting compounds in the production chain of organic chemicals. Whole families of organic chemicals were derived from them. When coal gave way to oil, ethylene and benzene replaced these as starting compounds. A comparable transition can be envisaged in respect to a shift from oil to biomass. In this case, the starting compounds of the future would be hydrogen, carbon monoxide, synthesis gas, methanol, furfural, glucose and adipic acid. The major difference with earlier transitions is that this transition is primarily motivated by environmental considerations that are not yet fully reflected in prices. The need for the transition is therefore not yet signalled in the marketplace and this leads to a delay in starting up the necessary work on R&D. Nonetheless, the downstream implications for the chemicals sector of a shift to biomass-derived starting compounds are significant. Wholly new synthesis routes will need to be explored and optimised if today's products are to be made sustainably from renewable feedstock.

The work of the STD programme in the area of chemicals and industrial materials shows that practically all of the major commodity products of the synthetic organic industrial chemicals sector can be produced, in principle, from plant materials, albeit that the chemical routes to each product will be altered strongly. The basic technologies exist. Proof of the existence of feasible conversions and step-by step sequences of conversions can be demonstrated for all of the top ten products. The aromatic compounds, especially phenol, will present the greatest problem, as these structures are not well represented in biomass. Nonetheless, solutions can be found for phenol production and substitute products can also be developed. Methanol emerges as a potentially versatile starting compound for producing the major lower olefins. Thus, all three of the new 'back-to-basics' technologies—HTU, flash pyrolysis and advanced gasification—offer ways of making bulk commodity products. Of greatest importance, all three offer production routes to bulk commodity products from biomass of any type and not just from selected parts of biomass, such as starches or sugars. In addition, there are opportunities for augmenting the flow of useful biomass raw materials for use in chemical production by specifically developing industrial crop varieties that can be grown under marginal conditions.

In respect to all of these opportunities, R&D efforts need to be refocused to give more attention to technologies at the front end of the materials chain in the winning and primary processing of raw materials. The barriers to taking up the possibilities relate mostly to cost. These can be overcome by a combination of improvements in biomass supply conditions and biomass-processing technologies on the one hand and, on the other, by policy- or market-driven changes in the availability and costs of fossil raw materials. The tangible outcome of the STD work includes an agenda for technological R&D on the most promising back-to-basics technologies, which is being carried forward by chemical companies and technical research institutes within the Netherlands. Related to this is a programme of research into the economic conditions and logistical/organisational innovations that would be required to facilitate ecological restructuring in the organic chemicals and fuels sector.

Chapter 11
Strategic niche management: the case of the mobile hydrogen fuel cell

This chapter is based on the work of members of the STD project on transportation and sustainable mobility and draws particularly on the work of Philip Vergragt and Diane van Noort (Vergragt and van Noort 1996). It extends the discussion begun in Chapters 9 and 10 about the need to wean our societies and economies away from using fossil-derived fuels. Especially it focuses on finding replacements for today's transport technologies, which include gasoline as the dominant energy carrier and the internal combustion engine as the dominant technology for converting the energy contained within gasoline to the energy service 'mobility'. Today's transportation technologies are unsustainable and the 'well-to-wheel' energy chain is very inefficient. In deciding which alternative technological directions to pursue, several needs must be taken into account. These extend beyond the straightforward need for sustainable mobility to include, also, the needs to achieve a sustainable energy system and to secure a smooth transition to sustainability from today's unsustainable transport and energy arrangements. The STD transport project evaluated potential alternative fuel and vehicle-propulsion technologies and drew out the implications for the Netherlands of contributing to the development of a hydrogen-based transport system using hydrogen-rich fuels and hydrogen fuel cells, which are among the most promising technological options.

The analysis is framed against the desire to minimise mobility loss and disruption cost in the transition to a future sustainable transport system, which should be based on renewable energy sources, clean energy currencies and efficient energy conversion technologies. The analysis also reflects the Dutch context. Although there is no major Dutch vehicle-manufacturing industry, there is a substantial petrochemicals sector based on adding value to imported oil feedstock.

Dutch industry plays a major international role in the production, conversion and distribution of liquid fuels, which are the dominant energy currencies in today's transport system. The future of the transportation energy market and of the factors and forces affecting it are therefore of considerable importance to the Dutch society and economy. Worldwide there is also a huge investment in a distribution network for liquid energy carriers, which is estimated to be of the order of US$200 billion. A final feature of the analysis was its concern to identify and define a strategic niche within which hydrogen technologies could be explored, developed, introduced and refined in a relatively protected context.[1] Since hydrogen-based transport technologies will not be competitive with fossil-based technologies for some time to come, at least in mass-market contexts, their development cannot be on the basis of open-market competition. Reserving a market share for a new technology protects it from technologies already well advanced along their learning curves of cost reduction and performance improvement and gives the practical experience needed to move the new technology along its own learning curve.

STRATEGIC PROBLEM ORIENTATION

We can set out the major requirements for a sustainable future transport system quite easily, although translating these into operational definitions is more diffi-cult. The major requirements are that the transport system should meet mobility needs at the same time as meeting environmental, economic, social and geo-political compatibility constraints. In addition, costs of all types should be mini-

1   The concept of strategic niche management is already well established and the approach has already been used to promote environmentally compatible technologies in several earlier contexts. In the US, for example, government procurement policy specifies a minimum recycled fibre content of stationery bought for government use. Because government is a major consumer of stationery, this policy secures a minimum market share for recycled fibre and paper products produced from it. This protected market niche is sufficiently large and guaranteed for it to stimulate the pulp and paper sector to develop improved technologies for fibre recovery and re-use even when the products may, in the beginning, be more expensive and have lower performance than products based on primary fibre. Legislation enacted by the State of California that requires vehicle suppliers to supply an accelerating share of zero-emission vehicles is another example. The Californian market is too large for manufacturers and suppliers to ignore. The legislation therefore pushes the pace of technological innovation by making innovation a condition for entry to the whole market, by safeguarding a part of the market for zero-emission vehicles and by giving notice that the proportion of the market given over exclusively to zero-emission vehicles will increase over time. In the Nether-lands a similar approach has been used to stimulate development and innovation of wind turbine technology by guaranteeing the purchase of renewable electricity at a price equivalent to the cost of gas-fired peak production.

mised in moving from today's arrangements to whatever future arrangements are proposed. Determining mobility need is difficult if not impossible. The best that can be done is probably to set out the different options that can be taken to reduce the environmental burden of transport on both the demand and supply sides of the equation and seek to define operational principles by which progress might be made on each front. In terms of compatibility constraints in respect to remaining demands, it is easier to define operational principles.[2] A sustainable transport system should be based on renewable energy sources. It should be based on inherently non-polluting technologies. It should also have highly efficient final conversion technologies. This is needed because the requirement to use renewable energy sources inevitably introduces inefficiencies at the front end of the source-to-service energy chain, which must be counterbalanced further along the chain by high efficiency in the final conversion of energy to mobility. Against this backdrop, we can evaluate current arrangements in the transport system to identify the major problems.

Our present societies and economies are characterised by high levels of mobility facilitated by well-developed road networks and high levels of vehicle ownership. Gasoline-driven internal combustion engines and diesel engines power most of the vehicles. Land use patterns and transport demands have co-evolved alongside the transport supply arrangements so that the capacity to be highly mobile and the demand for mobility have been mutually supportive. Current arrangements have become deeply entrenched and represent a coherent and internally consistent 'systems solution'. In terms of technology, the most important elements of the system are vehicles, vehicle-propulsion technologies, the physical transportation infrastructure, the technologies for traffic control and the technologies for fuel winning, refining and distribution. The dominant vehicle-propulsion technology— the internal combustion engine—is positioned at the interface between these. Moreover, it is the pivotal technology for the whole transport systems solution and has enormous upstream influence over the whole energy system. This is because each energy-service technology is designed to work with a specific final energy currency. Because the internal combustion engine (ICE) uses gasoline and because demand for mobility has grown strongly on the back of automobile technology, the demand for gasoline and hence oil has also grown strongly.

---

2   These have been set out already by analysts such as Quakernaat (1995) and Rogner (1998). To be environmentally compatible, the fluxes to and from the future energy system should be coherent with natural energy and material fluxes and they should not perturb natural equilibria. The energy system needs also to be socio-politically acceptable in terms of convenience, level of risk, economic affordability, supply security and concern over nuclear proliferation. Notwithstanding technological uncertainty, these compatibility constraints hold rather robust implications for the choice of elements within the energy system architecture because they limit the choice of energy sources, currencies and conversion technologies (Rogner 1998).

Today's transport systems solution is effective. However, it is far from sustainable or efficient in its use of energy. Transport energy use accounts for more than one-quarter of worldwide commercial energy consumption and is the fastest-growing energy end-use category. The average growth in transport energy use has been 2.7% per annum over the past 25 years. Virtually all of the energy is from non-renewable sources, some of which are also insecure.[3] Transport is responsible for 30% of all energy-related anthropogenic carbon dioxide (CO_2) emissions. The most significant transport sources are passenger cars (3.6 billion tonnes of CO_2 annually) and trucks (2.4 billion tonnes of CO_2 annually).[4] Both the passenger car and truck contributions to CO_2 emissions are individually greater than the energy-related emissions of any single industrial sector. Today's transport technologies are also associated with the production of oxides of nitrogen (NO_x)[5] and sulphur dioxide (SO_2). There are also significant land requirements for road infrastructures.

The issues involved in reducing the environmental impact in the transport sector are complex, partly because of the reciprocal linkage between the demand for and supply of transport.[6] In principle, there are at least four different broad solution directions. One lies in demand management, which includes a wide range of

3   The major resource is oil, which cannot last indefinitely. The largest sources of supply are not necessarily secure. Current exceptions to oil-based transport arrangements are electrically powered public transport systems, such as railway, tramway and road-based electric vehicles when electricity is produced from renewable sources, and Brazil's vehicle fleet, which is powered by biomass-derived ethanol fuel.

4   Transport energy end-use is increasing faster within industrialised countries than within the developing world. Another disturbing trend is that the energy intensity of goods transport is increasing in industrialised societies, such as the Netherlands, owing to shifts to faster and/or more powerful vehicles.

5   It is important to draw a distinction between nitrogen-based pollutants and all these others. NO_x is formed not by the combustion of fuels *per se* (since nitrogen is not a component of fuels), but by contact between air (which contains both nitrogen and oxygen) and high-temperature engine surfaces. At very high temperatures, nitrogen in the air can be oxidised to produce a range of compounds with potentially harmful environmental and health effects. In principle, the problem of NO_x pollution can be avoided if fuels are burned or otherwise transformed to deliver energy services at temperatures below 1,000°C.

6   There are considerable problems in defining transport need. To some extent, transport is a derived demand. The scale and pattern of demand is related to the level and structure of production and consumption in the economy and to the absolute and relative locations of people, places and activities. It is also a function of the willingness to pay for the trip and its costs. Some potential transport demands are more cost-sensitive than are others. One of the most important factors is how easy or difficult it is to avoid the journey, to link trips into a single (multi-purpose) journey or to find a substitute for physical transportation, whether of people or of goods. Journeys can be avoided by relocation, changes in land uses, or greater mixing of land uses. New technologies, such as flexible manufacturing and information technologies, also offer opportunities to avoid the need for physical movement of people and goods: for example, by offering economies of scope over economies of scale, providing facility for remote working or substituting electronic information transmission for transport of hard copy.

measures that would lead to a reduction in the need or demand for the physical movement of goods and people. Another lies in changing modal splits so that fewer vehicle-kilometres are needed to fulfil the same transport demand and the most sustainable mode is used for each transport task. A third lies in increasing vehicle loads and optimising routes so that fewer environmental resources are needed to meet the same transport task. The fourth lies in changing the technical characteristics of transportation technologies such that these are intrinsically more sustainable, such as by developing non-polluting vehicles powered by renewable energy and by employing information technology to smooth traffic flows and increase the efficiency with which road space is used.

Some argue that reduction in the amount of transport ought to be a strategic goal and that transport demand is a measure of the inefficiency of logistical arrangements and the economy. Certainly, because many components of transport cost are external to the private decision calculus, many journeys that deliver net private benefits impose net social costs. These journeys are inefficient in terms of resource allocation.[7] On the other hand, there are some physical transport needs that are essential, many journeys that deliver net social benefits and many situations where physical transport is the most efficient of all alternative logistical arrangements. A further complication is that, for many people, mobility is an important element in a high quality of life. Making a journey may even be an end in itself. Another consideration is that transport is associated with pronounced economies of scale and diminishing marginal costs owing to the high fixed cost of infrastructure. Until congestion sets in, average and marginal costs of individual journeys can reduce as network use rises. Increasing mobility has a profound impact on lowering production costs economy-wide and means that it is difficult, and maybe meaningless, to look at the costs and benefits of journeys in isolation. There may be positive as well as negative externalities from individual journeys.[8]

All four broad solution directions are likely to feature in the transition to a sustainable transport system. For the STD transport project, however, it was most important to focus on sustainable transport technology. There are limits to the extent of any mobility reductions or modal shifts likely to be deemed feasible,

---

7 Patterns of transport demand in Western industrialised societies have co-evolved with patterns of land use under market conditions where natural resource productivity has never been a priority, where road-building has been publicly financed, and where levels of (demand-led) road provision have been economically and environmentally excessive. Under these conditions, where many of the social and environmental costs of transport are not borne by the trip makers themselves but are 'externalised', patterns of land use and transport demand may represent an excessive use of transport.

8 The low cost of transport may have important economy-wide spin-offs, especially in lowering unit production costs and in increasing international competitiveness. There is also a high level of fixed investment of financial and environmental capital already sunk in the physical infrastructure of roads and buildings, whose value would fall if mobility were to be reduced.

desirable or acceptable within our societies and economies.[9] By implication, there will always be a substantial residual demand for transport. Also, the need for demand reduction will be the lesser (and so, also, will be the disruption to lifestyles, physical infrastructures and economic activities) if fundamentally more sustainable transport technologies can be developed. A focus on technology is also in keeping with the STD objectives of stimulating technological innovation and evaluating the extent to which technology can contribute to sustainability. The starting point for the project, therefore, was the 'need' to develop sustainable, clean and safe technologies in respect to uncertain but potentially large future demands for mobility.[10] In this, special emphasis was placed on vehicle-propulsion technology.[11]

THE IMPORTANCE OF VEHICLE-PROPULSION TECHNOLOGY

One reason for the poor environmental performance of the transport system is the low well-to-wheel energy efficiency achieved by today's transport technologies. The refining and distribution of gasoline from the primary source (oil) is quite efficient (80%). However, the conversion of gasoline to useful transportation service (per-

9   Unnecessary, economically unwarranted and easily substitutable transport demands would be sensitive to full-cost pricing. Land use patterns are also sensitive to transport cost. At the other extreme, however, no physical land use arrangement can overcome the heterogeneity of space. Some degree of separation of functionally related places and activities is inevitable. Nor can immaterial alternatives to physical movement be found for every transport demand.

10   The 'need' was interpreted as to facilitate the highest possible level of mobility consistent with doing so sustainably and efficiently. This has the advantage of potentially minimising the disruption and transition costs in moving to a sustainable solution, minimising opposition from vested interests and providing important stakeholders with incentives to seek technological solutions. This approach is fundamentally different from one that would begin by seeking to estimate a minimum level of essential transport need and aim at reducing demand to this level. Once sustainable technologies are identified, a level of transport demand that might be catered for sustainably can be determined. This is not to argue against the use of demand management measures. Indeed, instruments to reduce demand in the near and medium terms (such as moves toward full-cost pricing in the road transport sector) could be designed also to provide incentives for R&D on sustainable transport technology.

11   The project pursued three lines of research in this direction only one of which—that on vehicle-propulsion systems—is reported in this chapter. The two lines of research not considered here were concerned with reducing the high demand for physical space made by our transport systems, especially for sensitive space in the heart of cities. One line looked at novel uses of information technology to manage traffic flows more efficiently, which would also have spin-offs for fuel-related eco-efficiency concerns. The other looked at possibilities for removing freight traffic from surface roads by developing alternative underground freight moving systems in specially designed, dedicated, low-friction, trackless tunnels. An illustration project to test such a system has been started in the context of a freight link between wholesale markets and Schipol airport.

sonal mobility) is very low. Over 81% of the energy content of onboard gasoline is lost as waste heat. Less than 19% is actually converted into vehicle movement. Of this, 4.2% is used to overcome rolling resistance, 10.0% is used to overcome air resistance and 4.3% is used to accelerate or to climb. Of this useful energy, the bulk (17%) is harnessed to move the weight of the car. Only 2% is expended in moving the weight of its occupants (Fussler with James 1996). These figures clearly demonstrate that, in exergy terms, the well-to-wheel energy efficiency of the transport energy chain is at best 16% and, depending on how far the analysis is carried, may be below 2%. It also demonstrates that, in common with most final energy services, the minimum energy requirement of the service 'mobility' is commensurately low. A final observation is that the greatest loss of useful energy in the chain occurs when converting the energy carrier gasoline to mobility. In effect, the final energy service equipment, the vehicle-mounted ICE, contributes most to the overall inefficiency of the well-to-wheel energy chain.

The clear consequence of inefficiency in source-to-service energy chains is that there is a corresponding scope for improving environmental performance by increasing the energy efficiency. Equally, the greatest scope for performance improvement lies at the point in the chain where inefficiency is greatest. In terms of the transport system, the greatest potential for narrowing the inefficiency gap lies in improving the energy end-use technologies and infrastructures since these are currently responsible for the greatest losses in useful energy (Rogner 1998). However, there are immutable physical limits to the efficiency improvement potential of today's internal combustion engine technology precisely because it is a heat engine. Moreover, ICE technology has been incrementally improved already for a period of almost 100 years and the existing designs are close to the theoretical efficiency limits. In spite of considerable efforts made to improve efficiencies, the average fuel efficiency of production-line ICE vehicles has improved by less than 1% annually over the past 20 years. Some analysts, for example Claude Fussler, consider adherence to ICE technology to be one of the most egregious examples of 'innovation lethargy' (Fussler with James 1996).[12] By implication, to increase the efficiency of the transport source-to-service chain as a whole, there is a need to develop an alternative technology to the ICE.

The urgency of revisiting the question of vehicle propulsion is compounded because of the added inefficiencies that would be introduced inevitably by a shift to renewable energy. The ultimate renewable energy source is solar power, which is inherently a decentralised, dispersed, low-density energy source that must be captured over large areas, entailing considerable expense in building appropriate infrastructures, and then concentrated and converted into useful forms. The specific energy density of sunlight constitutes an immutable constraint whether

12   Fussler and James write that: 'Average fuel efficiency is stuck, and stuck on very low' (Fussler with James 1996: 6).

the energy is captured directly using photovoltaic cells or indirectly in the form of biomass. In its direct form, solar energy is also an intermittent power source with a supply pattern in space and time that is poorly matched to energy demand. With the exception of biomass, solar energy is also locked into electricity production and electricity can only be stored in limited quantities. Converting electricity into storable chemical energy introduces yet further inefficiencies and expense. Since upstream inefficiency cannot be avoided when dealing with renewable sources, it must be counterbalanced by efficiency gains elsewhere if sustainable energy chains are to become competitive. Again, this argues strongly for R&D on alternatives to ICE technology.[13]

ENVISIONING A TARGET TRANSPORT SYSTEM

A future vision of a sustainable transport system that is based on renewable energy sources and is inherently non-polluting holds implications for the technological architecture of the entire transport system.[14] The requirement to use only renewable energy sources limits the choice of primary energy to solar, geothermal, tidal and nuclear energy. Solar energy, which is received on Earth in the form of transmitted electromagnetic radiation, can be subdivided into two classes: direct and indirect. Direct sources include radiant heat and light; indirect sources include biomass and the energies of movement, such as hydro, wind and wave power. The

13   Vehicle-propulsion technology is by far the most critical issue in transport technology development because it embraces issues not only of vehicle engineering, but also of fuel supply. It also takes us to the heart of the strategic importance of the transport sector for achieving sustainability economy-wide. The significance of vehicle-propulsion technology is that it lies at the interface between the transport and the energy systems. One reason why the internal combustion engine has had so profound an effect on 20th-century development and its unsustainability is that developments in the energy sector are largely driven by the requirements of energy end-use conversion technologies. The demand for energy products is a function of the demand for energy services and the characteristics of the final energy conversion technologies that deliver services from the energy (Rogner 1998). Analysts of energy systems point out that, while the flow of energy through the energy system goes from source to service, the energy system is service-driven from the bottom up. Downstream market conditions largely determine the upstream situation. The introduction of the gasoline-driven ICE can be seen as a defining moment for the developmental trajectory of the 20th century. In similar fashion, a new vehicle-propulsion technology could be equally pivotal for the development trajectory of the 21st century.
14   Such a vision also suggests diagnostics that could be used to set targets and to monitor progress toward sustainability. Operational indicators of progress could include: the renewable fraction of primary energy sources, the C:H ratio of the primary energy mix, the hydrogen and electricity fraction of final currencies, the fraction of electrical, electrochemical and catalytic energy conversion and the overall energy efficiency of source-to-service chains.

need to replace fossil sources is reinforced because of the even more stringent requirement for a quasi zero-pollution energy system, which requires the elimination of fossil carbon. Only carbon-neutral sources and technologies are indicated (Box 11.1). Sustainably managed biomass is the only non-fossil carbon-rich energy source. In turn, any restriction to only biomass sources of carbon implies a drastic overall decarbonisation of the energy system with downstream implications both for currencies and for conversion technologies.

The only clean energy currencies are heat, electricity, hydrogen, and biomass-derived bio-fuels such as ethanol and methanol. The cleanest and most efficient use of bio-fuels involves a reforming step to hydrogen before final conversion. This leaves electricity and hydrogen as the two principal candidates as final energy currencies. Combustion and heat engines as technologies for converting energy are counter-indicated in all but very controlled conditions. Clearly, oxidation of fossil-sourced carbon is counter-indicated in any case. But, while combustion could operate on renewable fuels, combustion with air as the oxidant produces nitrogen compounds and is, in any case, inefficient. Electrical, electrochemical or catalytic conversion technologies, unlike combustion, are efficient and pollution-free at the point of conversion. The shift to using renewable sources of primary energy also holds implications for the choice of final conversion technology. Using renewable energy inevitably reduces efficiency at the front end of the source-to-service chain, which has to be offset later in the chain using highly efficient end-service technologies. This underscores the need for a shift from using heat/combustion engines to using electrical, electrochemical and catalytic conversion technologies for delivering final energy services. Electrochemical conversion technologies are typically more than twice as efficient as ICE technologies when converting final energy currencies into mobility.

In sum, in order to be compatible with a sustainable energy future, any future vehicle-propulsion technology will need to be highly efficient in converting its final energy currency to mobility. The final energy currency could be hydrogen or electricity. Both are clean at point of delivery, both can be produced from renewable sources on a sustainable basis and both can be converted efficiently to mobility. More especially, however, the two energy currencies are complementary. A sustainable energy system will depend on both currencies being used as complements to one another. There are three main reasons for this. The first is that most renewable sources of energy are locked into electricity production and electricity cannot easily be stored. In order to store renewable energy and to even out demand and supply, it will be necessary to convert renewable electricity into a storable chemical energy. Hydrogen provides a means for doing this. By using hydrogen as an energy buffer, the variable and intermittent yield of the renewable energy sources that generate electricity can be converted into a constant flow of energy. This is indispensable for post-fossil energy management (Quakernaat 1995). The second and related reason is that not only can electricity be converted into

1. WHY POLLUTION IS A MORE URGENT CONCERN THAN DEPLETION

Currently, total world primary commercial energy consumption is around 350 EJ. The structure of world primary supply is: oil/petroleum (39%), coal (29%), natural gas (20%), nuclear power (6%), and other alternatives such as hydro-power, wind power and biomass (8%). In summary, fossil energy is dominant today. But, in respect to restructuring the energy system, depletion of fossil reserves is not the dominant concern. Since pre-industrial times, the concentration of CO_2 in the atmosphere has risen from 280 to 350 parts per million (ppm), mostly as a consequence of burning fossil fuels. The cumulative use of carbon by humanity to date amounts to 230 gigatonnes (Gt). Stabilising the atmospheric concentration of CO_2 over the next century at a level no more than double the pre-industrial concentration (450 ppm) would involve containing cumulative emissions from all sources over the period 1990–2100 to below 650 GtC, equal to an average annual emission rate of 5.7–5.9 GtC (IPCC 1996a, 1996b). This annual emission target is *less than today's emission level* of 7.4 GtC for all sources and 6.0 GtC for energy sources and *substantially below trend extrapolations of future emission levels*. The IPCC reference level for annual emissions from energy sources is 20 GtC/yr by 2100 and for cumulative emissions over the period 1990–2100 is 1,500 GtC (IPCC 1994).

The total carbon content of all identified and inferred conventional oil, gas and coal occurrences has been estimated to exceed 20,000 Gt (IIASA/WEC 1995). Taking an energy price of $20, $30 and $40 per barrel oil equivalent (pboe) as the reference price for economic recoverability and using a dynamic model to account for technical changes in the exploration and extraction situation (reserves, costs, etc.), the carbon content of recoverable reserves is estimated to be 2,000, 3,000 and 7,000 Gt at each respective price level (Rogner 1990, 1997; Nakicenovic *et al.* 1993). Thus, at a conservative energy price of only $20 (not much different from historical norms), it would be economically viable to recover reserves with a cumulative carbon content of 2,000 Gt. Such a level is an order of magnitude greater than the amount of carbon already released to the atmosphere over the whole of the industrial era to date. If the 200–300 Gt of cumulative carbon emissions to date give reason for severe concern about climatic stability, it is reasonable to conclude that most of the indicated fossil reserves should remain untapped in the ground. Against this backdrop, fossil energy resource scarcity is unlikely to be the driver for future resource shifts; on the contrary, the abundance of cheap fossil energy is likely to delay eco-transition in the energy sector, which may therefore need to be policy-led.

2. WHY A LONG-TERM SHIFT TO RENEWABLE ENERGY IS NEEDED

Today, oil is the main primary energy, followed by coal and gas. The oil and coal shares are, however, declining relative to gas and nuclear energy, which are both increasing. These shifts in the structure of primary energy supply are consistent with long-term trends toward using progressively higher-quality (higher energy density) sources. Biomass (wood) was the first dominant source. In sequence, wood gave way to coal and coal to oil. Now natural gas is set to take over from oil to become the dominant primary fuel. Because of the different C:H ratios of primary energies, these historical inter-source shifts can be interpreted also as intermolecular substitutions of hydrogen and carbon molecules (Rogner 1998). The shift from wood (C:H ratio 0.7) to coal (C:H ratio 1.0) increased the C:H ratio during the first industrial revolution. Since the 1920s, however, the C:H ratio of the primary energy supply has decreased progressively at an average annual rate of 0.2%, slowing up only temporarily after the mid-1970s as a result of oil policies adopted by OECD countries in the wake of the oil crisis. This decarbonisation of the primary energy supply is likely to continue as oil (C:H ratio 0.5) and coal (C:H ratio 1.0) are replaced by natural gas (C:H ratio 0.25) and carbon-free nuclear energy. Today, the C:H ratio of primary energy sources worldwide is 0.6, which is equal to 0.015 tonnes of carbon per GJ (tC/GJ).

Global energy consumption is on a rapidly rising trajectory. Assuming unchanged policies and continued trends in the size and structure of populations and economic activities, ➡

Box 11.1 **Energy restructuring: three critical arguments**

→   energy consumption by 2100 would increase seven-fold to more than 2,000 EJ (Gouse *et al.* 1992). But no fossil energy system, even one based totally on the most benign of fossil fuels, natural gas, can have a C:H ratio of less than 0.25; i.e. 0.0075 tC/GJ. With a world population of 10 billion people by 2050 and a corresponding increase in energy service demand, a C:H ratio of 0.25 may well be insufficient for the long-term target of stabilising atmospheric greenhouse gas concentrations. The carbon consequences of an energy demand over 800–900 EJ—a demand that now seems inevitable even with stringent and successful demand management and efficiency measures—would be too great to stabilise atmospheric greenhouse gas concentrations even if met totally by natural gas. Any reduction below a C:H ratio of 0.25, however, ultimately requires the use of non-fossil energy sources. By implication, 50 years from now (when demand for energy service is anticipated to be several times greater than it is today), renewable energy sources must have taken over a substantial part of worldwide energy consumption.

3. WHY FOSSIL FUEL DECARBONISATION IS A NECESSARY INTERIM STRATEGY

However, since clean and sustainable energy chains are too expensive to implement at present, global energy management will have to contend with a large number of non-sustainable energy supply structures during a transitional period (Quakernaat 1995). For several decades to come, developing countries, especially, will depend on fossil sources and on coal in particular. The key problem area for richer countries is in relation to transportation fuels. In the IPCC reference scenario, the major contributor to growing carbon emissions over the period to 2100 is the production of synthetic fuel from coal. Unless countervailing measures are put in place, the quantity of coal used for this purpose by 2100 is projected to be more than four times today's total coal use. A reasonable conclusion for the transition period until clean fuel can be produced from renewable sources is that 'the major greenhouse challenge will be to avoid enormous increase in CO_2 emissions arising from the production of synthetic fuels from coal' (Williams 1998:181). During a transition period, then, fossil energy must be decarbonised before use to avoid further build-up of atmospheric concentrations of greenhouse gases.

SOME CONCLUSIONS AND IMPLICATIONS FOR RESTRUCTURING THE ENERGY SYSTEM

The situation calls for a multi-track agenda for restructuring the energy system. The main strategies—on a continuing and worldwide basis—are demand limitation, energy efficiency improvement and economy-wide reduction in the energy and materials intensities of product chains. Much can be achieved by reducing the energy and materials intensities of production–consumption chains and by shifting toward concepts of prosperity that are different from those prevailing today, assessing future demand for energy on the basis of service needs rather than on trend extrapolation (Reddy and Goldemberg 1990; Hueting *et al.* 1992; Quakernaat and Vermij 1994). These strategies need to be supported in the short and medium terms with actions to decarbonise fossil energy in developed and developing countries and by efforts, in today's industrialised countries, to reduce the cost and increase the performance of technologies forming part of longer-term sustainable energy solutions (Quakernaat 1995).

Box 11.1 (continued)

hydrogen but that this conversion can also be reversed. Third, the two carriers are fundamentally different. They have different characteristics that make them uniquely appropriate for use in specific applications.

As to this last point, the comparative advantages of hydrogen are that it can be used to store energy in any quantity, can act as an energy buffer and can also be used as a chemical or material feedstock. In contrast, only electricity can be used to process, transmit and store information. Such differences render the two currencies suitable for performing different energy tasks in a future sustainable economy. The differences are extremely important in the case of transport. Although batteries can be used to store electricity and provide power to independent vehicles on the road system, they have limited storage capacities and power density. There is also a substantial recharging time. These limit the range of battery-powered vehicles and also the effectiveness with which they can deliver final transport services. For all but low-power energy requirements, transport-service technologies that use electricity as final currency must be hard-wired to the electricity production and distribution system. The practical consequence is that electricity is most suitable for rail-based systems where a connection to the grid can be established. In turn, this means that electricity is most appropriate for fixed-route traffic, such as to deliver public transport services. Electric-powered battery vehicles are viable, but they are likely to be less efficient and effective than hydrogen-based vehicles for independent mobility needs.

THE CURRENT STATUS OF HYDROGEN

Hydrogen and electricity are both secondary energy carriers. They are not found on Earth as fossil fuels are found. Both have to be produced using relatively large amounts of energy at high cost (Johansson *et al.* 1993). Hydrogen can be produced from natural gas, oil, coal and biomass using chemical conversion technologies.[15] It can also be obtained by electrolysis of water, which releases oxygen as a by-product. In effect, electrolysis of water allows electrical energy to be converted to chemical energy and stored in the form of hydrogen molecules. The electricity for water electrolysis, just as electricity for any other purpose, can be generated from various sources. It can be generated from fossil fuel, nuclear fuel or renewably as

15 Hydrogen is released during the digestion of organic substances. However, because other reaction products such as ethanol and acetate are also generated in this process, gasification of biomass is a better option for the production of hydrogen. This gasification technology is applied today *inter alia* to wood and other vegetable residues such as rice chaff. For the production of hydrogen on an industrial scale, the quantity and quality of the biomass feedstock would need to be assured: for example, by targeted energy cropping (Quakernaat 1995).

SOURCE	PRODUCTION PROCESS	PRODUCTION COST (1994) ECU¢/kWh		PROJECTED COST (2020) ECU¢/kWh
Hydro-electricity		2.5 – 5.0		
Wind electricity	Water electrolysis	7.5 – 15.0		3.7–6.0
Solar parabolic trough		15.0 – 20.0		‹10.0
Off-peak nuclear		5.0 – 7.0		
Natural gas	Steam reforming	4.6 – 6.0		
Other fossil fuels	Partial oxidation of oil; coal gasification + steam reforming			
Fossil fuel and electricity	The Kvaerner process (with carbon black as by-product)			
Biomass	Biomass conversion		4.5	2.5

Table 11.1 **Hydrogen production routes and costs**

Source: Adapted from Wurster and Zittel 1994

photovoltaic, wind and hydro-electricity. Complementary processes can reverse the outcome by converting hydrogen to electricity. Heat from hydrogen combustion can generate steam to drive a turbine. More efficiently, electricity can be produced directly from hydrogen through electrochemical conversion, which is the reverse of water electrolysis. As long ago as 1839, William Grove, a British physicist, demonstrated that the electrochemical union of hydrogen and oxygen generates electricity and, as a by-product, water.[16]

The key links in hydrogen energy chains are production, conditioning, transport, storage, distribution and final conversion to energy service. For each link, there are several alternative technologies. As a commercial product, hydrogen can be produced using a range of sources and processes. In addition, hydrogen is a by-product of several chemical processes. Table 11.1 presents illustrative production routes, processes and costs. For large-scale hydrogen production, today's cheapest routes (after production as a chemical by-product) are electrolytic production from large-scale hydro-electricity, steam reforming of natural gas and conversion of biomass. Although for most production routes hydrogen production costs far exceed those of fossil fuels, hydrogen produced via electrolysis from off-peak electricity can already be produced today at a cost that is competitive with liquid

16   In a fuel cell, hydrogen and oxygen are combined without combustion to deliver electricity. Producing electricity in a fuel cell closes the cycle of the initial substance, water.

natural gas as a fuel for mobile applications (Wurster and Zittel 1994). The production cost of hydrogen from natural gas is also low enough for this to be competitive with the pump price of gasoline (taxes included) on an equivalent energy-content basis (US30–40¢ per litre).

Hydrogen is a gas under ambient conditions so it must be transformed into a transportable, storable and dispensable form after production: a process known as conditioning. Hydrogen can be conditioned as a compressed gas. Alternatively, it can be conditioned as a liquid at cryogenic temperatures and transported and stored in insulated tanks. A third possibility is to condition hydrogen as a hydride, a chemical compound capable of binding hydrogen. Hydride compounds can be liquid or metallic. Liquid hydrides include methanol and methyl-cyclohexane. Metal hydrides are metal alloys that offer space for hydrogen atoms within their lattices, which are 'charged' with pressurised hydrogen. Equally, sponge iron (Fe_3O_4) can be used to 'store' hydrogen. The injection of hydrogen into sponge iron effects a reduction of Fe_3O_4 to Fe, at the same time liberating water vapour. The stored chemical energy can be re-released by injecting water vapour, which returns the Fe to Fe_3O_4, at the same time liberating hydrogen. Another option is cryo-adsorption. This uses a material porous to hydrogen, such as activated carbon fibre, which is charged with pressurised gaseous hydrogen at low temperatures. The storage density depends on the temperature and pressure conditions (Table 11.2).[17]

The global production of hydrogen is currently around 500 billion m³ annually.[18] There is a global liquefaction capacity of around 200 tonnes/day. Several countries have dedicated industrial hydrogen pipeline networks together amounting to several hundred kilometres. Storage is mostly as a compressed gas or a liquid. In sum, hydrogen is used routinely today without problem, but only in restricted and expert domains in industry, where it is used as an intermediate material.[19] Only in

17   These forms of conditioning differ in respect to cost, safety and convenience. Other important factors in deciding the appropriate form of conditioning are the quantities of hydrogen involved, transport distance, length of storage, ultimate use and purity require-ment. For instance, when large volumes of hydrogen have to be transported over long distances, hydrogen should be compressed and transported via pipelines or liquefied for bulk transportation by tanker at cryogenic temperatures. If hydrogen is to be produced electrolytically, then one possibility is local decentralised production and conditioning close to the point of final use. Mobile applications make specific demands in terms of safety, the ease and speed of refuelling, the energy density of onboard hydrogen storage and the weight and volume of any equipment needed for fuel handling and conversion.

18   The main production routes are via catalytic steam reforming of natural gas and naphtha, the partial oxidation of hydrocarbons, pyrolysis of coal and crude oil, gasoline reforming, dehydrogenation of hydrocarbons and organic compounds and water elec-trolysis. In addition, hydrogen is produced as a by-product in the chlor-alkali industry.

19   With one exception—the use of liquid hydrogen as a rocket fuel—the use of hydrogen is today confined to industrial applications. Half of total production is used in non-energetic applications, mostly for ammonia synthesis (23%) or for crude oil processing (22%). In respect to energetic applications, 30% of all hydrogen is used to provide

Form of conditioning	CHARACTERISTICS	IMPLICATIONS FOR TRANSPORT/ STORAGE	IMPLICATIONS FOR MOBILE APPLICATIONS
GASEOUS HYDROGEN	*Pipeline pressures:* 0.4 MPa—local 2.0 MPa—intermediate 6.0 MPa—long distance *Storage pressure:* 20–30 MPa *Hydrogen density at storage pressure:* 25 gH/litre Compressors, valves and gaskets of today's natural gas pipelines need to be adapted.	Good for pressurised pipeline transport at 0.4–6.0 MPa. Energy densities too low for distribution of containerised compressed gas to be economic over more than short distances. Could be produced locally by electrolysis at decentralised filling stations and conditioned to 20–30 MPa.	Offers higher on-vehicle energy storage density than either metal hydride or sponge iron. High-pressure onboard storage containers needed. Prototype containers under development (30 MPa) using plastic composite materials and metal liners.
LIQUID HYDROGEN	Hydrogen storage density: 80 gH/litre. Evaporation causes losses and pressure build-up in containment vessels. Evaporation rate of 1%–3% daily from 100–400 litre tanks.	Suitable for trans-continental shipment by tanker/barge and for truck delivery using super-insulated containers. Suitable for direct use as a vehicle fuel but new dispensing equipment needed.	Cryogenic hoses and nozzles needed that can be disconnected in cryogenic state to allow rapid refuelling of successive vehicles. Improved cryogenic onboard tanks needed. First trials with vacuum super-insulated tanks at 0.4 MPa and 20K.
LIQUID HYDRIDES	Are liquid, stable and safe under ambient conditions. High hydrogen and energy density.	For distribution, storage and vehicle refuelling can be treated like gasoline. Suitable for pipeline distribution and long-term storage.	Onboard equipment for dehydrogenation adds to vehicle weight. Hydrogen carrier substance may have to be recycled.
METAL HYDRIDES	Storage pressure: 5 MPa Energy density lower than sponge iron Very safe	Heat is liberated during charging and must be supplied for discharging.	Vehicle applications need low-/medium-temperature hydrides as only low-grade waste heat available. First trials with MH tanks at 5 MPa
SPONGE IRON	Energy density higher than metal hydrides. Low investment cost: ECU1.5/kWh.	Currently an order of magnitude cheaper than competitors.	Requires steam to be injected into the storage to liberate hydrogen.
CRYO-ADSORPTION	Energy density at 3.5 MPa and 77K: 25 gH/l		First trials with cryo-adsorption vessels in US: 4–6 MPa and 60–150 K

Table 11.2 **Considerations in hydrogen conditioning**

Source: Adapted from Wurster and Zittel 1994

the very narrow domain of aerospace applications is it used today as a fuel. Nonetheless, in principle, hydrogen has a potentially important general role to play in a future sustainable energy system and, because of its capacity to work alongside renewable electricity as an energy store and buffer, it has a very specific role in a future sustainable transport system. For hydrogen to be extended from industrial uses into applications in the wider public domain, the whole hydrogen chain from production through to delivery of final energy services must be technically feasible and economically, environmentally and socio-politically acceptable. The principal barrier is cost (Quakernaat 1995).[20]

SUSTAINABLE HYDROGEN PRODUCTION

For a transport system based on hydrogen fuel to be sustainable, the hydrogen should be produced from renewable primary sources. The efficiency of sustainable hydrogen production depends on the primary energy source and the subsequent conversion steps in the source-to-service pathway. Whatever the production route, however, the efficiency of sustainable hydrogen production is always lower than the efficiency of gasoline production and distribution, which is 80% from oil well to currency-onboard. If hydrogen is produced electrolytically, the production efficiency depends on the initial efficiency of electricity production as well as on the efficiency of water electrolysis and hydrogen conditioning. The production efficiency of nuclear electricity is around 40%–50%. For solar electricity captured using photovoltaic cells, it can be considered to be 100% if solar energy is regarded as 'free'. The subsequent hydrogen pathways, including electrolytic production, liquefaction, distribution and dispensing, incur a loss of around half of the useful energy. Overall, then, the 'source to currency-on-vehicle' efficiency is 20%–25% for hydrogen produced from nuclear sources and 50% for hydrogen produced from photovoltaic electricity. The non-electrolytic route for hydrogen production from

---

process heat and 20% is used indirectly in the petroleum and synfuel industries (BMFT [Bundesministerium für Forschung und Technologie] cited in Wurster and Zittel 1994; SWWS [Solare Wasserstoffwirtschaft Bonn] cited in Wurster and Zittel 1994).

20 Technological progress, especially in biomass, wind and solar technologies, could reduce hydrogen production costs and improve production efficiencies significantly in the coming years. Although water electrolysis offers opportunity for carbon-free production of hydrogen, progress depends on the development of less expensive and more efficient equipment. Large-scale solar (photovoltaic) hydrogen will only be attainable when solar cells are cheaper and production efficiencies much higher than now. Practical trials with solar electrolysis are taking place in Germany (Winter and Fuchs 1991; Barra and Coiante 1993). Other relevant factors are the prices of alternative energy carriers and the willingness of society to pay a premium for carbon-poor energy (Quakernaat 1995).

biomass via methanol with on-vehicle reformation to hydrogen offers a source to currency-on-vehicle efficiency of 45%–55%.

This efficiency gap between gasoline and hydrogen fuel production could be substantially redressed farther down the energy chain if, rather than using hydrogen to fuel combustion engines, customised energy-service technologies developed specifically for hydrogen fuels are used instead. Catalytic and electrochemical conversion technologies, which convert hydrogen directly into electricity without first burning it, offer higher fuel efficiency than combustion engines. For transport applications, fuel-cell electric drive concepts are two to three times more efficient than internal combustion engines with mechanical transmission systems. Achievable fuel efficiencies from currency-on-vehicle to useful wheel traction are 20% for vehicles with internal combustion engines but 45%–55% for electric drive vehicles powered by fuel cells using methanol or hydrogen fuel. These give an overall source-to-service chain efficiency of 16% for gasoline-fuelled vehicles with combustion engines, but up to 28% for hydrogen-fuelled fuel-cell vehicles with electric drives. In effect, the lower production efficiency of hydrogen can be offset by the higher efficiency of fuel conversion to mobility, enabling hydrogen fuel-cell technologies to compete on cost and efficiency criteria with today's gasoline-ICE technology on a source-to-service basis (Rogner 1998).

BACKCASTING AND THE MANAGEMENT OF TRANSITION

Backcasting from this future vision of an energy system based on sustainable sources and clean currencies, it becomes important to imagine how the transition from today's energy system might be arranged. Today's energy system is based on carbon-rich energy sources. In cases such as transport, the carbon-rich energy source (oil) is converted to a chemical energy currency (gasoline), which is used directly within the conversion technology (the internal combustion engine) to deliver the required energy service (propulsion) at point of need. Under today's arrangements, 'cleaning' vehicle emissions at the end of the vehicle exhaust pipe is out of the question, even though this would be the only way to prevent loss of CO_2 to the atmosphere. There is nowhere to store the CO_2 that is produced from thousands of decentralised mobile sources—even temporarily—and no way of moving it to safe, centralised, long-term storage. Also, the costs of CO_2 scrubbing would be high since CO_2 represents only 13% of vehicle exhaust gases.

A more practical alternative to 'flue' gas cleaning would be 'fuel' gas cleaning: that is, the removal of carbon from fuels before their use (Williams 1998). Moreover, as well as contributing to a cleaner energy system in the short and medium terms, fuel gas cleaning could play an important role in achieving a smooth transition toward a long-term sustainable energy future. A carbon-rich feedstock can be reformed to a hydrogen-rich feedstock by a relatively simple process, the water–gas shift reaction (Box 11.2). This involves transferring the chemical energy of carbon to hydrogen. The energy is 'shifted' through a three-step process that also allows the carbon of the original feedstock to be captured in the form of CO_2 during the final step. The efficiency of this process is quite high—between 60% and 85% depending on which carbon-rich feedstock is being converted. Because fuel gas cleaning takes place centrally, unlike flue gas cleaning, the captured CO_2 is then available to be stored so that it does not enter the atmosphere and interfere with atmospheric processes. This raises the possibility of sequestering the CO_2 in some secure reservoir. One possibility is to use exhausted natural gas fields as storage sites for captured CO_2. A spin-off is that this puts the gas fields under pressure once again and forces out remaining natural gas. The additional natural gas production is estimated to offset much of the cost of CO_2 capture and storage.

Fuel decarbonisation with carbon sequestration may be important for the prospects of managing a smooth transition to a long-term energy future based on sustainable sources and clean currencies because the production of hydrogen-rich fuels is integral to the process. Fuel decarbonisation could thus play a role in a sequenced progression away from the highly polluting use of fossil fuels, through

Box 11.2 **Fuel gas cleaning: the water–gas shift reaction**

THE FIRST STEP IN THE CONVERSION IS TO GASIFY THE CARBON (C) BY PARTIAL oxidation to carbon monoxide (CO). The CO retains 71.9% of the original energy of C. (The higher heating value of C is 393 MJ/mol. The higher heating value of CO is 283 MJ/mol.) The 'shift' of the bulk of energy occurs in the second step, when CO is reacted with steam to give a gaseous mixture of carbon dioxide (CO_2) and hydrogen (H_2). This is also an exothermic reaction. The energy released (41.2 MJ/mol) is just less than that needed to raise the required steam (44.0 MJ/mol). The small amount of make-up energy (2.8 MJ/mol) can be met using heat obtained in step one (110.5 MJ/mol). The third step involves separating the produced H_2 from the gaseous H_2/CO_2 mixture. This can be achieved by pressure swing adsorption (PSA). PSA makes use of the ability of some porous materials to adsorb specific molecules selectively at high pressure and to desorb them once pressure is lowered. Separation is achieved by cyclically raising and lowering (swinging) the pressure. PSA can recover 90% of the produced H_2 at up to 99.999% purity (Williams 1998).

The theoretical efficiency of generating 1 mole of H_2 from 1 mole of C by the shift reaction is 72.6% over the first two process steps. The third, separation, step reduces the efficiency to about 65% over the entire process. Because different feedstocks have different chemical compositions, the process steps and the efficiency of the energy shift depend on the precise reaction conditions. Efficiencies are highest (84%) when converting natural gas (CH_4) because there is no need for an oxidation step. This process begins directly by reacting natural gas with steam. Efficiencies for coal ($\sim CH_{0.8}O_{0.08}$) via a process of oxygen-blown gasification and for biomass ($\sim CH_{1.5}O_{0.7}$) via steam oxidation are slightly lower at around 64% each (Williams *et al.* 1995).

the manufacture of cleaner (non-polluting) fuels from fossil sources and, ultimately, to the manufacture of clean fuels from renewable (non-fossil) sources. Each step along the way marks a reduction in carbon emission and an increase in hydrogen use. Indeed, the progression could end with the possibility of delivering energy services while achieving a net removal of carbon from the atmosphere. This would be possible if hydrogen were produced from biomass with carbon sequestration and storage.

It has been suggested that this could provide a technical means by which countries committed to reducing atmospheric concentrations of CO_2 could compensate for others that are either unwilling or unable to cut their own CO_2 emissions. More important, however, is that fuel gas cleaning with sequestration provides a way to use fossil fuels cleanly during a transition period, which extends the time available for making the transition and makes it easier for more countries to begin the transition earlier. While not sustainable in the long term, this may provide an important near-term stepping stone, especially for countries without access to natural gas. In this context, several analysts have noted that the transition to using natural gas seen over the past decade in most Western economies is consistent with progress toward sustainability since this represents a shift in energy sources toward those with a lower C:H ratio. It also represents a shift toward using gaseous energy carriers (Lee *et al.* 1988).[21] Fuel gas cleaning with CO_2 sequestration would mark a further step in the transition to using gaseous, hydrogen-rich energy currencies.

EVALUATION OF THE TECHNOLOGICAL ALTERNATIVES

Against this backdrop, members of the STD transport and mobility project team drew up a shortlist of promising vehicle-propulsion systems. In drawing up the shortlist, project members considered three clusters of internal vehicle-propulsion systems, some hybrid systems based on combinations of battery and fuel cell technologies and several external vehicle-propulsion systems. Fourteen different external systems were evaluated using Delphi-style iterative questionnaire surveys of experts in the field of vehicle design. However, all of the external systems were found to offer lower chances of successful development and fewer spin-offs in the event of successful development than any internal propulsion system (Box 11.3). Accordingly, project members concentrated on internal propulsion technologies in

---

21   More generally, a gas pipeline infrastructure and gas-handling expertise are essential to a sustainable energy future (Lee *et al.* 1988) while infrastructures developed to handle natural gas could, in principle, be adapted for the future transport of hydrogen produced from sustainable sources.

VEHICLE-PROPULSION TECHNOLOGIES CAN BE DIVIDED BETWEEN THOSE WHERE the energy supply is carried onboard (internal) and those where energy is supplied to the vehicle from outside (external). Another potential distinction is between vehicles that do and do not carry the equipment onboard to convert energy to movement. Although today's cars carry both an onboard fuel tank and engine, there are potential advantages in external energy supply and/or external traction. If energy is provided externally, such as in the case of electric trains today, energy production can take place remotely. This makes it possible to generate power on a larger scale, which may offer higher efficiency, lower emissions and lower exposure to emissions. Also, the choice of primary fuel is more flexible while the sources of that fuel can be varied. Moreover, vehicles without fuel are lighter. The argument can be carried a step further if there is no need for an onboard engine. The drive system for ICE vehicles accounts for 20%–25% of the total vehicle weight. In the case of electric vehicles, the drive system and battery represent over 50% of the total vehicle weight. The dead weight of equipment contributes significantly to overall source-to-service chain energy inefficiency.

The STD transport team organised surveys of vehicle experts to list possible external energy supply and/or vehicle-propulsion technologies. For example:

▷ Vehicles with a battery for use in local traffic but with a facility to couple to overhead electricity lines or ground-based cables when on main roads

▷ Electric vehicles with a flywheel or battery recharged frequently at stops/crossroads

▷ Electrical recharging of batteries while moving, using wireless energy transmission

▷ Electrical recharging of batteries by vehicle-mounted solar cells

▷ A linear electric motor where the vehicle is rotor and the track is stator

▷ Towing lines or conveyors to which a vehicle can hitch when on main roads.

Most of these technologies are concerned with providing power to electric vehicles for direct use or to recharge batteries. The systems differ in respect to whether power is transferred when vehicles are moving or stationary and whether or not the transfer is hard-wired. Other technologies involve hybrid systems, which extend the range of battery-powered vehicles by providing a facility to hitch onto a towing or carrying device when on main roads.

These different options were evaluated through a Delphi process involving an international panel of experts drawn from different countries and institutional backgrounds. The aim was to clarify the feasibility of each technology, any expected contribution to a more sustainable transport system, the expected timing of introduction and the nature of the main obstacles to be overcome. Most of the potential technologies were considered to have major drawbacks or to be unworkable except in unrealistic or niche operating conditions. In every case, for private transport applications the panel consensus was that there are better alternative technologies with internal power supply offering fewer barriers to implementation, higher overall eco-efficiency, higher safety and lower cost.

For illustrative purposes, comments in respect to specific technologies include:

▷ **Electric battery vehicles supplied on main roads by overhead or ground-based cables.** The basic idea is promising because it would enable electric vehicles to travel longer distances. The main obstacle is infrastructure, which would be complicated, expensive and unsightly. The infrastructure costs would be prohibitive for a niche market (small demand) and the operational problems would be unmanageable for a general market (mass demand). The limits to mobility and costs make this technology more suitable for large vehicles on set routes, such as buses and trucks. The costs are likely to be higher than those involved in developing the batteries or the fuel cells that would make such systems unnecessary. →

Box 11.3 **External vehicle-propulsion systems**

▷ **Electric vehicles with a flywheel or a battery recharged at stops such as crossroads.** There are already some commercial realisations of flywheel technology, but only with low-speed fly-wheels (Oerlikon bus, Parry tram, NMVB Gent). The development of low-weight flywheels at reasonable cost is, however, a problem. The flywheels, bearings and vacuum-pumping systems needed would be expensive due to the sophisticated production processes. Reliability is a problem. In earlier experiments, problems arose with buses unable to reach charging points and getting stuck in traffic. Compared with batteries, there are few advantages. Comparable power can be delivered by a number of battery technologies. Only the power in–out efficiency of flywheels is better. The infrastructure for recharging is more sophisticated than for batteries and, therefore, more expensive. There has also been a commercial realisation of fast and fre-quent battery recharging—the *Akkutriebwagen* of the German railways. The trial was sus-pended owing to high costs and low technical performance. The recharge time, energy storage capacity, battery weight and range are all technical problems. The life-span of the batteries is reduced by the high charging power.

▷ **Electrical recharging of batteries by vehicle-mounted solar cells.** The power density of panels mounted on a vehicle (up to 100 W/m²) cannot generate a significant portion of the energy required for driving. It therefore makes little sense to mount solar cells directly on vehicles. While the use of renewable energy is a major goal, this should be done in the most cost-effective way. Panels should be exposed to sunlight for as long as possible and be oriented correctly. This makes solar cells more suitable for mounting on fixed surfaces, such as roofs. Solar cells have a longer life-span than cars, which also makes them more suitable for use on buildings.

The conclusion drawn from the panel's comments was of support for electric vehicles, but not of technologies for external energy supply to them or for external propulsion. The pref-erence was for electric vehicles that are entirely internally propelled. In effect, if batteries and fuel cells can be improved there will be no need for external energy supply or propulsion. Only if these technologies do not meet expectations did panellists suggest that an external energy supply technology could become viable in the future. In this case external energy trans-mission to moving vehicles via overhead pantographs could be useful in limited applications contexts.

Box 11.3 (continued)

which both the energy currency and the conversion technology are carried on-vehicle (Box 11.4). While it is too early to discount any of the internal propulsion options, those based on hydrogen or hydrogen-rich fuels and fuel-cell technologies offer the greatest long-term potential to compete effectively on cost, performance, efficiency and sustainability criteria. Moreover, hydrogen fuel-cell technology is strategically important for supporting long-term restructuring toward a sustain-able energy system and enabling carbon-based fossil fuels to be phased out gradually. The STD transport project therefore focused on fuel cells as a strategic, potentially pivotal technology for the 21st century as an alternative or complement to battery technologies.

BATTERY-POWERED ELECTRIC VEHICLES

Battery powered vehicles are potentially sustainable and clean since electricity can be produced from renewable sources and electric engines produce no pollution at the point of energy delivery. Electrical engines are more efficient than internal combustion engines. However, recharging is slow, vehicle range is restricted and the battery is heavy, which reduces efficiency. The drive system and battery represent over 50% of the total weight of today's battery cars.

BIO-FUEL-POWERED ICE VEHICLES

Bio-fuels, such as biomass-derived ethanol or methanol, are renewable and carbon-neutral. If burned directly in internal combustion engines, however, the high temperatures produce NO_x. Also, the efficiency of the fuel use is substantially lower than could be achieved by first reforming the fuel to hydrogen and then converting hydrogen electrochemically in a specifically designed fuel cell. The energy efficiency of a source-to-service chain involving bio-fuels and ICEs would be too low to offset the inefficiencies introduced at the front end of the chain in producing the bio-fuel.

METHANOL- OR HYDROGEN-FUELLED FUEL-CELL VEHICLES

A hydrogen fuel-cell vehicle offers the prospect of combining the advantages and avoiding the disadvantages of both the battery-powered and bio-fuel-ICE alternatives. A fuel-cell vehicle could deliver more than twice the fuel economy of an internal combustion engine powered by bio-fuel, avoid NO_x production and represent a carbon-neutral solution or better. Compared with a battery vehicle, a hydrogen fuel-cell vehicle would face no significant refuelling delays and have no restrictions on use or range.

HYBRID BATTERY AND FUEL-CELL VEHICLES

Combining batteries and fuel cells may overcome some of the technical problems with fuel cells, especially in the early stages of commercialisation. Batteries could augment power output during acceleration and climbing and provide power while the fuel cell is warming up. This solution would also reduce costs, because a smaller stack of fuel cells would be needed.

Box 11.4 **Internal vehicle-propulsion systems**

**HYDROGEN
FUEL CELLS**

As already mentioned, fuel cells are not a new technology. The basic principle on which they operate was discovered over 150 years ago and working cells have been implemented already in practical applications, such as to provide power and water for spacecraft. The basic principle of a fuel cell is straightforward. A fuel cell consists of two electrodes separated by an electrolyte that allows the passage of protons but not electrons. At the anode, hydrogen fuel is split into protons and electrons. The protons pass through the electrolyte to the cathode, while the electrons travel through an external circuit to the cathode. Oxygen is supplied at the cathode. The hydrogen ions recombine with the oxygen to produce water at the cathode; the chemical energy of the hydrogen is converted to electricity. The best-known application was by NASA in the 1950s and 1960s in the Gemini space programme. Nonetheless, formidable technical and economic obstacles currently bar practical applications in more mundane contexts such as road transportation.

The technical obstacles lie in three broad areas. First is the question of fuel production. Second is the problem of securing a fuel supply to the fuel cell. Most hydrogen storage technologies involve heavy and bulky equipment. Liquid fuels rich in hydrogen—such as methanol—offer a more manageable storage solution, but present problems of their own. They have to be reformed to hydrogen onboard the vehicle, just before use. Extracting the hydrogen is a difficult chemical process that must be catalysed over platinum. The conversion equipment takes up space and adds further to the dead weight, reducing the efficiency of the overall source-to-service energy chain. Onboard reformation of methanol can also cause carbon contamination of the produced fuel, which poisons cell components. Third, the power density of all hydrogen fuel-cell technologies is still too low to be practical for transport applications. Production costs are also too high. The general consensus of experts in the field is that for an electrochemical engine to become competitive, production costs need to be reduced to around \$50–200 per kilowatt, which represents a cost reduction of at least a factor of 100 over today's production costs. These technical obstacles are formidable. At the moment, hydrogen fuel-cell technology is a long way from being competitive on performance or cost criteria with more conventional technologies. Nonetheless, the potential advantages that fuel cells offer are substantial, especially over stand-alone ICE or battery technologies, which makes them an interesting technology to explore for the long term.

One of the major advantages is that fuel cells, unlike batteries, do not have to be recharged. The difference between a fuel cell and a battery is that in a fuel cell the reactants are the hydrogen fuel and oxygen, whereas in a battery the materials used for the electrodes are the reactants. A battery is therefore simply an energy storage device, which must be charged up before discharge. Since recharging takes time and there is a limit to how much energy can be stored in the electrodes, a battery cannot be used to supply a continuous stream of electricity. A battery is merely an energy buffer. Its major use is to enable the input and output streams of electricity to be de-coupled physically and in time. In contrast, a fuel cell is an energy conversion device that can produce a continuous stream of electricity so long as continuous streams of hydrogen fuel and oxygen are supplied at its electrodes. Unlike recharging of a battery, refuelling of a vehicle fuel tank can be achieved in minutes. A fuel cell-equipped vehicle can therefore be driven continuously without any range restriction. Other advantages in mobile applications are that fuel cells are quiet, can be designed for any desired power output and are highly energy-efficient. The theoretically achievable chemical-to-electrical energy conversion efficiency of a fuel cell is very high, but is reduced in practice by the limited rate at which oxygen can be made to react with hydrogen at the cathode. Nonetheless, with good catalysts, the achievable efficiency is 45%–60%, which is twice the optimum efficiency achievable with an ICE and three times the average ICE efficiency.[22]

Fuel cells can be designed to operate at different temperatures. The only fuel cells of interest for mobile applications are ones that operate at low temperatures. This avoids NO_x production and allows aqueous materials to be used, which is important for some designs. However, the need for low operating temperatures creates a technical problem since the electrochemical reactions need to be catalysed at low temperatures in order to generate a useful flow of electricity. The catalyst at both electrodes is platinum, which is a scarce material and very expensive.[23] Another problem is that the electrical resistance of the electrolyte restricts the fuel cell efficiency. To overcome this, a strongly acid or alkaline electrolyte can be used, as these have lower electrical resistance. This gives rise to a choice of candidate fuel-cell technologies for use in mobile applications differentiated on the basis of the material used for the electrolyte. The main candidates are alkaline fuel cells, acid fuel cells and synthetic polymer (proton-exchange membrane) fuel cells. Each has advantages and disadvantages.

At present, alkaline fuel cells must operate on industrial-quality hydrogen, as any carbon contamination of the hydrogen fuel will react with the electrolyte to form a solid carbonate. This rules out onboard reformation of methanol as a means of supplying fuel. By implication, alkaline fuel cells have to operate on compressed or liquefied hydrogen, which creates very difficult technical and logistical problems in arranging fuel delivery. Both the fuel storage equipment and the fuel cell are large and heavy. However, some progress is being made to reduce the sensitivity of alkaline fuel cells to carbon contamination and they offer cost advantages over other fuel cell designs, especially because they need less platinum catalyst than any other low-temperature fuel-cell technology.

Acid fuel cells use water to conduct hydrogen protons, so cells need to operate at low temperatures, which lowers efficiency. Aqueous acids are also volatile and unstable. Acid fuel cells also take a long time to reach their operating temperature, which limits their usefulness for applications in private cars. An exception to the low temperature requirement is the phosphoric acid fuel cell, which has a higher permissible operating temperature (over 200°C), which means that the waste heat can be used to generate steam for the onboard reformation of methanol to hydrogen. It also gives the fuel cell some resistance to carbon contamination from the resulting, slightly impure, hydrogen fuel. Even so, phosphoric acid fuel cells suffer from the same low power density and slow starting problems of other acid fuel cells.

22   Fuel cells do not have to idle when the vehicle is stationary, unlike ICEs.
23   The cost of platinum is around $10 per gram. Enormous progress has been made in reducing the platinum requirement. In 1985 the requirement for polymer fuel cells was over 15 grams per kilowatt of power. This has been reduced to 0.5 grams per kilowatt of power over the past 15 years, a more than thirty-fold reduction (Appleby 1999).

Synthetic polymer fuel cells get around the instability problem of aqueous acids by using a synthetic polymer membrane as the electrolyte. The membrane is made of an acid-containing material that allows the passage of protons. The best-known material is Nafion, made by DuPont. Originally, the polymer membranes were expensive and short-lived. Nafion, however, has a lifetime of 60,000 hours. The major problem is that synthetic polymer technology uses a lot of platinum catalyst per kilowatt of power generated, at least 0.5 grams per kilowatt. Since a car needs around 50 kW of power to accelerate, a car powered exclusively by a polymer fuel cell would need a substantial fuel cell stack. Even with the best technologies, the cost of the platinum catalyst would be prohibitive. Moreover, there would be sourcing problems if an entire car fleet were based on this technology. The platinum requirement of 0.5 grams per kilowatt is not generalisable, since it would take only 6 million 50 kW vehicles (only one-sixth of today's production) to use up the total world production of platinum (Appleby 1999). Also, this technology is not well suited for combining with onboard methanol reformation, as there is no obvious way of providing the necessary steam for the reformation process other than by burning some of the fuel. The carbon-sensitivity of the design would require a catalytic conversion step in the production of hydrogen fuel and the use of a platinum-ruthenium catalyst at the anode to prevent poisoning.

The proton-exchange membrane fuel cell is seen by many as a leading contender for use in cars, but the problem of the platinum requirement remains a significant obstacle. Although progress has been made in reducing the quantity of platinum used per kilowatt of power, a substitute for platinum has not been found. Clearly, each fuel cell option presents a set of technical and economic challenges. In turn, fuel cells are only one component of the overall vehicle-propulsion system. There are many other elements, including fuels, fuel storage systems, fuel reformation systems and energy conversion systems that can be combined in different ways. This means that no candidate technology stands out clearly from all others. Each has advantages and disadvantages. Some technological challenges are common to all designs, but others are more specific. Since these challenges may or may not be overcome by R&D, the ultimate outcome depends on the success in making technological breakthroughs of one sort or another: for example, in the field of catalysis or fuel reformation or stack design. Solutions might also be found by combining less powerful fuel cell stacks with batteries to give a hybrid vehicle that could offer continuous power generation, back-up power for acceleration, immediate starting and regenerative braking.

To date, three approaches have been followed in building practical demonstration vehicles.

> ▷ A straightforward proton-exchange membrane fuel cell using industrial-quality elemental hydrogen carried onboard the vehicle. For reasons already given, this technology could work for some, but not for all vehicles owing to the high platinum demand and the high cost implied.

▷ A hybrid design combining a battery and a phosphoric acid fuel cell. The battery is used to provide power for acceleration and is recharged during braking. This reduces the power requirement from the fuel cell to about 15 kW and, so, saves on the cost of platinum catalyst. This solution can use onboard reformation of methanol to produce the hydrogen fuel and, so, avoids the problems associated with using elemental hydrogen as the stored-onboard fuel.

▷ A hybrid design combining a battery and an alkaline fuel cell. The alkaline fuel cell uses only one-fifth as much platinum as the proton-exchange fuel cell (0.1 grams per kilowatt). At the moment, this approach depends on using industrial-grade hydrogen. However, some progress has been made on developing carbonate-insensitive cathodes, which potentially might open the way to using onboard reformation of methanol in conjunction with alkaline fuel-cell technology.

TOWARD AN R&D AGENDA

Fuel cells are at an early stage in their development and there is a range of possibilities for improving technical performance and reducing costs. One research objective is to reduce the volume and weight of the cells to improve power density. A second major research area is related to overcoming the problems of hydrogen production and storage. Another is to develop processes and equipment for on-demand generation of hydrogen from methanol. There are research questions relating to cell design and the cell manufacturing process. Today's cells are individually fabricated. Series production would significantly lower costs, but this depends on reaching a threshold level of market demand for units. To operate in a vehicle, the fuel cell has to be integrated into the overall vehicle design. In turn, a fuel-cell vehicle has to function as part of the wider transportation system and has to be integrated into the physical infrastructure. If, for example, the vehicle is designed to operate on stored hydrogen, this has implications for the fuel supply and distribution infrastructure. If, alternatively, the vehicle is designed to generate hydrogen onboard from methanol, there may be possibilities for adapting the present fuel supply and distribution system, including the pipeline network and the current infrastructure of filling stations. Either way, the implications for implementation need to be worked out.

In summary, the major R&D issues surrounding hydrogen energy chains for mobile applications are concerned with:

▷ Reducing the cost and improving the performance of hydrogen production

▷ Producing hydrogen in carbon-neutral ways or from renewable sources

▷ Minimising the life-cycle environmental impact of entire hydrogen source-to-service energy chains

▷ Improving the technologies of the supporting hydrogen infrastructure (i.e. the infrastructure for hydrogen conditioning, handling, transport, storage and distribution)

▷ Improving catalytic and electrochemical conversion

▷ Designing the geographical configuration of the hydrogen infrastructure

All of the questions involved need to be explored along the relevant economic, socio-political and technical acceptability criteria of cost, safety, environmental impact, operability and efficiency.

THE DUTCH
CONTEXT

Internationally, several research projects are under way to improve the performance and lower the costs of fuel cells. In a few places, demonstration projects are developing.[24] However, questions relating to the integration of hydrogen into society and the economy and to the management of transition of the energy system are less well researched. In the context of the Netherlands, these latter questions more closely match with Dutch interests and expertise. The Netherlands does not have a significant automobile industry and is unlikely to become a major producer of fuel cells in the future owing to the small size of the potential domestic market. However, the Netherlands plays a major role in the international liquid fuels market and has a sizeable natural gas industry. The Netherlands has substantial interests and expertise in the areas of fuel technology. By implication, it is particularly well placed to tackle those questions on the R&D agenda concerning the production and distribution of hydrogen-rich currencies and the handling of hydrogen fuel on vehicles.

Another important factor for the Netherlands is its own, long-term energy outlook. Today, the main energy sources for the Netherlands are oil and gas. In terms of the national energy balance, domestic gas production meets domestic needs and provides an exportable surplus to offset against oil imports. Netting all imports and exports of energy (which are substantial), the nation is broadly energy-self-sufficient today. This situation will change as the domestic natural gas sources in the North Sea dwindle over the coming decades. Hydrogen is therefore a long-

---

24 At present, Ballard in Canada is developing buses with polymer fuel cells and in the USA there is a programme for the development of a solid polymer fuel cell for cars. In the Eureka Fuel Cell demonstration project there is a bus fuelled by an alkaline fuel cell.

term energy option for the Netherlands. There are no insurmountable technological barriers to its development, but there are high costs involved in introducing hydrogen energy chains as replacements for natural gas and oil. There are also a number of technical and strategic considerations relating to implementation. Given the geographical situation of the Netherlands, hydrogen would have to be imported via tanker or pipeline from energy-surplus, carbon-free production sources abroad. Before then, hydrogen could be produced within the Netherlands from domestic and imported fossil fuels with carbon sequestration, making use of exhausted natural gas fields as storage reservoirs for sequestered CO_2. Both the technological and geographical configuration of any future hydrogen infrastructure should be explored. These are strategic issues since the optimal choice of technologies as well as the scale and degree of centralisation of facilities will depend on the extent to which hydrogen chains serving different end-uses are integrated.

STRATEGIC NICHE MANAGEMENT

Coherent 'systems' sketches of fuel cells, vehicles, infrastructures and fuel-supply systems are needed to probe these questions and to smooth a route toward implementation. The primary need is to agree on an illustrative project in the course of which key questions for R&D and implementation will emerge and can be addressed by stakeholders and actors. The illustrative project may ultimately involve the construction and testing of prototypes but, in the beginning, much can be achieved (and high costs avoided) by modelling and simulation. In the event of moving on to prototypes, a protected market niche needs to be found where experiments can be made with nascent technologies to provide feedback on technical performance, user experience and implementation issues.

Through discussions on the part of members of the innovation network established by the STD programme, an illustration project was proposed based on the design and ultimate construction of a ship powered by fuel cells. There were several reasons for this choice. While the Netherlands does not have its own automobile industry, it has a shipbuilding tradition and a reputation for innovation in ship design. Also, the Netherlands has a high level of ship traffic. It is a trading nation and its main port, Rotterdam, is one of the largest and busiest in the world. Owing to its strategic position within Europe, the Netherlands also serves the trading activities of its European neighbours, including Germany. The port is the hub of import, export and value-adding activities of both national and international significance. In a future sustainable economy, the port of Rotterdam would potentially be a European hub for hydrogen import, re-export and value-adding activities, much as it is today for oil and oil products.[25] Clean ships that can carry and be fuelled by hydrogen are important for the sustainability of ship-borne trade and for the economic future of the port. They are also important for the overall eco-efficiency of production chains of all traded commodities and goods manufactured from these.

Notwithstanding these advantages, the real environmental benefits would lie in the adoption of fuel cells in land transportation. But a ship can act as a first niche both to tackle an R&D agenda and to explore both implementation issues and the potential spin-offs surrounding hydrogen. Especially, a ship can provide an early laboratory in which to explore the practical problems faced in supplying hydrogen to onboard fuel cells and of integrating the fuel cells into the rest of the vehicle architecture as well as into the wider economy. A practical experiment using a ship would throw up the same technical and strategic questions for the fuel-supply industry that would be faced if the fuel cells were to be powering a land-based vehicle. However, these experiments could be performed more readily and sooner on a ship because experimentation does not depend on waiting for very compact cells to be developed.

CONCLUSION

By exploring the interrelationships between the transport and energy sectors, it was possible to pinpoint mobile hydrogen fuel cells as breakthrough technologies with economy-wide spin-off potential. The project was also able to support a constituency of stakeholders in defining an associated R&D agenda and in identifying a strategic niche within which to develop designs for a mobile hydrogen fuel cell-powered vehicle and probe implementation issues. Specifically, these relate to the technical implications of implementing a transportation technology based on a combination of a hydrogen fuel chain with a mobile hydrogen fuel cell propulsion technology. The palette of R&D options was reviewed especially in relation to the specific contribution that the Netherlands is positioned to make toward transition in the transport and energy systems. Specific R&D bottlenecks to be tackled were highlighted concerning the production and handling of hydrogen fuel, carbon sequestration and storage during hydrogen production from fossil fuels, and on-demand generation of hydrogen from methanol on vehicles.

25 The proposal to base an illustration project around a ship also integrates the STD work on vehicle-propulsion systems with that on a sustainable main port. Although the nature of trade and shipping will change in a future sustainable society, Rotterdam will remain as a major transportation hub, central to the Dutch economy. The STD programme addressed the issue of what sustainability implies for the function, design and operation of the main port in the future and for the organisational and managerial aspects of implementing strategic change. The development of more sustainable ship propulsion systems was one of five main needs identified for the make-over of the port.

In this project, the substantive conclusions are important as well as the methodological aspects. Both in the energy system and in energy-using applications, such as the provision of transport services, the possibilities to devise carbon-neutral and even carbon-negative energy chains are important for sustainability. In the long run, hydrogen can be derived from biomass. In the interim, it is also possible to derive hydrogen-rich fuels from fossil fuels. Importantly, CO_2 can be sequestered during this process by separation from the fuel gas. In principle, sequestered CO_2 could be stored in saline aquifers and exhausted natural gas fields at relatively low cost. The Netherlands is particularly well situated to develop this possibility because of its North Sea natural gas fields, which will gradually become exhausted over the next 20 years. Sequestration with CO_2 disposal means that it would be possible to produce hydrogen and methanol fuels even from coal with relatively low emission of CO_2 to the atmosphere. In combination with onboard reforming equipment and fuel cells, these interim technological possibilities may offer the prospect of a potentially seamless progression from today's transport arrangements to a more sustainable future.

There are also important innovation outcomes related to the take-up of these possibilities. The project successfully brought together a group of actors from different backgrounds whose efforts should be harmonised if a new transportation technology is to be implemented. The project also achieved consensus around a broad strategic direction, which is important for co-ordinating and harmonising R&D efforts. Joint working has led to a clear R&D agenda with tangible short-term research objectives related to the longer-term vision of a sustainable future transport system. Elements of this R&D agenda have already begun to be addressed by private- and public-sector actors. Several industrial actors in the Netherlands, for example, are now working on fuel systems for fuel cell applications in transport. Shell and DaimlerChrysler are working together on catalytic systems for on-demand conversion of methanol to hydrogen. Carbon sequestration is being followed up as a major research theme related to hydrogen production from fossil fuels. There are important spin-offs, too, in respect to technology transfer to countries now rapidly industrialising.[26]

---

26 The STD continuation programme is also in discussion with the Chinese government over a co-financed programme of R&D aimed at disseminating and adapting the STD methodology to Chinese conditions within case-study sectors. Preliminary discussions are focused on the energy and transport sectors as potential 'swing' sectors on which the basic orientation of general economic development within China might depend. This is an important outcome given the continuing strong rate of economic growth in China, the huge potential demand for energy and transport services and the fact that coal remains, at present and for the near-term future, the most abundant source of available energy for the Chinese economy and society. Clean coal technologies, although not a sustainable solution over the long term, may provide an important bridge toward a sustainable future in the context of current rapid economic growth in China.

Part 3

There is no difficulty in describing the STD programme, its rationale, underlying philosophy, methodology or case studies. However, there is an intrinsic difficulty in evaluating the programme and its contribution to sustainable technology development. Any tangible results of the programme in the form of new sustainable technologies will only emerge in the long term and the programme's influence on these will only ever be indirect and difficult to separate from other influences. Any commentary at this time must therefore be based on near-term and indirect indicators of influence—mostly related to the programme's influence over the goals and conduct of innovation processes—and be based on how well the programme has achieved the effectively intermediate objectives set out in Chapter 4. In this final chapter, our comments are structured around these interim objectives. Has the programme demonstrated that sustainable technologies are, in principle, possible? Was the programme able to develop and demonstrate an effective methodology for influencing innovation processes? Has the programme contributed to achieving wider societal appreciation of the sustainability imperative? Has the programme been able to initiate new and self-maintaining innovation trajectories? Are there continuation activities? Has the programme been able to leverage its own impact through a demonstration effect? As to the last of these questions, it is important that several countries have maintained a monitoring brief on the STD programme with a view to setting up comparable activities. The most systematic independent monitoring of the programme has been done on behalf of the German Parliament (Schramm and Wehling 1997).[1]

---

1   The German evaluation of the STD programme was made after a preliminary review of several national programmes, including ones in Sweden, Japan and the US.

ARE SUSTAINABLE
TECHNOLOGIES POSSIBLE?

One of the programme objectives was to demonstrate that sustainable technologies are possible in principle. The programme clearly shows that, in some fields of need, technologies with substantially improved resource productivity can be developed. In the need field of nutrition, protein foods that impose less than one-thirtieth of the environmental strain of pork are, in principle, feasible. In the field of textile cleaning, solutions are possible that use only a fraction of the resources used today. In the need field of mobility, energy chains based on renewable sources and carbon-neutral technologies are, in principle, possible. Overall, the programme identified and described a wide set of individual technologies that could contribute to providing eco-efficient solutions. More importantly, the programme was able to identify synergies among clusters of technologies, which combined could provide solutions offering eco-efficiency improvements of a factor 10 or more in some key areas of need.[2] Synergistic technology clusters were highlighted in scenarios and future visions in several application contexts.[3] Technologies that feature frequently in such scenarios can be identified as pivotal for sustainable development. Examples include photovoltaic technologies, sensor technologies and technologies based on exploiting the catalytic properties of some materials.

More especially, the various case studies illustrate the many ways in which technology can help to make better use of available eco-capacity. New knowledge can 'create' new resources, enable abundant and benign resources to substitute for those that are scarce or whose use poses problems and make existing resources more productive. Several case studies demonstrate ways of augmenting the resource base by exploiting hitherto under-exploited resources, such as solar energy, biomass and genetic information. Others demonstrate how new knowledge about plants, animals, energy and materials could make the existing resource base more productive: for example, by bringing deserts, brackish soils, mountainous regions and previously neglected plant and animal species into more productive use. New knowledge about basic properties of materials or specific types of material, such as chromophores, can open new ways of meeting needs that will allow renewable and biodegradable resources to substitute for traditional mineral resources.

---

2   The prospect for major improvement in the need field of nutrition is a case in point. The STD programme identified solution directions relating to clusters of technologies for producing unconventional foods while, at the same time, identifying clusters of technologies that could break through many long-standing constraints on conventional food production linked to soil, water and topography.

3   An example is provided by a combination of PAR-transparent photovoltaic technologies, fibre-optic cables and sensor technologies in the development of closed horticultural systems. Another is given by the combination of hydro-thermal upgrading, advanced gasification, flash pyrolysis, and C_1 technologies in developing production routes for organic chemicals from biomass feedstock.

Equally, new knowledge about how to process materials can enable an abundant resource or a waste to substitute for a scarce resource, such as might be achieved by water and pressure treatment of softwood to produce a hardwood substitute or by hydro-thermally upgrading sewage sludge to produce bio-crude.

The case studies also demonstrate how a systematic search for sources of waste and eco-inefficiency in present arrangements can be used to highlight improvement opportunities. The ways in which environmental resources are used today reflects past assumptions about the abundance and low value of resources, even though these assumptions are increasingly invalid. Today's approaches to resource and waste management are correspondingly crude. Biomass, water, energy and nutrients are all used inefficiently today. In the agricultural sector only about 10% of produced biomass is valorised. In the water and energy sectors, resources are wasted by failing to differentiate need.[4] Similarly, many active chemical agents used to deliver services, such as chemical fertilisers, pesticides, herbicides and detergents, are administered by blanket application today rather than by targeting accurately according to need. Today's failure to separate wastes of different type and quality also leads to eco-inefficiency. Separating materials from mixed waste-streams requires additional resources, while mixing can prevent valorisation of various components of the waste-stream altogether.[5] In all these cases, inefficient use of resources increases the draw on limited eco-capacity at both ends of the materials chain. The enormous eco-inefficiencies involved provide huge scope for eco-efficiency improvement through integral use of resources, resource cascading, resource targeting and by holding wastes of different qualities separately.

Nature is perhaps the best model for sustainable solutions.[6] Most anthropogenic technologies pale in comparison with the effectiveness, efficiency and sustainability of the 'technologies' applied in nature. The performance of natural systems, ecosystems and organisms can therefore offer benchmark standards against which the eco-efficiency of anthropogenic technologies can be compared as well as insights into how improvement might be achieved.[7] In many cases, the superior

4   Cascading water and energy down a usage hierarchy can substantially improve eco-efficiency.

5   Allowing small quantities of heavy metals to join large-volume waste-water streams, for example, delivers double disbenefits. The entire volume of waste-water must be treated to remove the contamination, however small the contamination is. Moreover, since treatment can never be 100% effective, some contamination will remain. Thus the whole of the organic content of the waste-water stream is rendered unusable for valorisation as a source of nutrients (see Chapter 7).

6   If proof of the existence of sustainable technology is needed, we need look no further than nature itself, which surrounds us with working examples, such as the 'metabolisms' of the planetary system, its component ecosystems and food chains, all of which operate on sustainable lines using renewable energy and closed material cycles.

7   The silkworm, for example, produces a single polymer thread, which has one of the highest strength:weight ratios of any known material, natural or man-made. The single fibre is up to 2 km long. It is synthesised at ambient temperature from a renewable mulberry leaf feedstock that provides for all the worm's energy and material require-

eco-efficiency of organisms can be exploited directly to provide needed functions, as is the case with bio-catalysts and bio-products.[8] Within the STD programme's illustrative case studies, this can be seen in the approach to using microorganisms in small-scale bio-refineries to convert solid biomass to liquid and gaseous energy carriers or to using bio-catalysts in chemicals production and task-specific enzymes to clean textiles. In other cases, nature's solutions can provide design analogues for technologies.[9] An important lesson is that eco-efficiency can be both a driver for technological improvement and a source of inspiration and ideas about improvement possibilities.

IS IT POSSIBLE TO INFLUENCE INNOVATION PROCESSES?

The case studies also show that it is possible to influence innovation processes in ways that improve the chances for sustainable technology development and diffusion. In several illustration projects, the methodology developed by the programme was effective in reorienting innovation efforts, making sustainability an innovation goal, integrating eco-efficiency criteria into technology design processes and building mechanisms to enable technological, cultural and structural innovations to be

ments. Studies of the silkworm show that the grub is capable of synthesising many complex proteins. The worm is now the subject of focused scientific study and is being used to produce other products, such as human growth hormone

8   Silkworms are not the only grub whose metabolising properties have attracted the attention of scientists. Recent work on insecticides, which began as an attempt to test the usefulness of viral infections for pest control purposes, led to the observation that cabbage looper caterpillars (*Trichoplusiani*) infected with specific viruses are capable of producing a wide range of different protein antibodies in their attempts to fight off disease. The proteins produced by the caterpillar are similar to several drugs used to control human diseases. By altering the virus to which the caterpillars are exposed, using genetic engineering techniques, the caterpillars make different proteins and can be induced to produce several commercially useful products. It has been suggested that the work opens the way for caterpillars to be used directly to produce drugs in the future. On another level, it reveals the enormous gap between the eco-efficiency of metabolic processes in organisms and the eco-efficiency of conventional technologies.

9   An example is how designers of robots have learned to minimise the information-processing power needed for robots to obtain information about their surrounding environment and to translate this into controls of robot movement by emulating the ways in which insects receive and process information. The compound eyes of insects give a two-dimensional image of the world around them. Information on the third dimension is provided only when insects move from side to side during flight. The combination of compound eyes and side-to-side movement minimises the processing power and therefore the size of the brain needed for gaining a three-dimensional image of the external environment. This is important for insects, which need to be compact and light in order to fly. It is important also for reducing the programming complexity for robot designs.

considered jointly. The programme was able to develop and demonstrate an effective methodology, which, although not incorporating any radically new elements, represents a first attempt to systematise a generalisable, coherent methodological approach to sustainable technology development. The method uses several elements and tools (most notably Constructive Technology Assessment and backcasting), adapts these to the specific applications context of sustainable technology development, and links them together to provide a sequenced, stepwise approach to innovation. A major achievement has been the creation of several new and now self-standing innovation networks with shared goals and agreed action plans for sustainable technology development. Another success has been to bring sustainability and its implications for technology and innovation to wide attention among members of Dutch society.

CONTINUATION ACTIVITIES

Many of the most promising ideas for new technological trajectories initiated by the STD programme through its case study projects have become embedded in continuation activities in the Netherlands and no longer receive any further support from the STD programme. Most are being carried forward through private funding arrangements. Several initiatives begun during the case study on fine chemicals are being progressed by Akzo-Nobel and DSM. Two Dutch companies working together with international collaborators are carrying forward the work on closed-system horticulture. Two companies, Stork and Holec, are working to downsize gas turbine technology to a scale compatible with applications in private cars using bio-methanol as a fuel. A foundation has been established and funded by corporate members of the STD C_1 chemistry network specifically for the purpose of carrying forward the work programme developed during the course of the case study. Financial support is being delivered by Shell and the Dutch electricity supply industry. The research agenda developed by the STD programme for the development of novel protein foods has been entirely adopted by one of the Dutch technology institutes, which is receiving funding to carry forward the research from Unilever, Gist-brocades and André de Haan. Further to ideas developed within the framework of the STD transport and mobility case study, a demonstration system for an underground freight transport solution to serve Schipol airport is to be developed. A consulting firm, RUPS, has been established with private-sector finance to help tackle the technical problems that the project encounters.

For new innovation trajectories that have a public-sector component—such as those dealing with water, land use and housing—the STD project has helped to establish new consortia of public and private bodies to carry the work forward. In

Almelo several different organisations concerned with fresh-water production, fresh-water distribution and waste-water treatment have made an agreement surrounding a long-term strategy for the development of the local water system along the lines worked out in the STD case study of the municipal water chain. The agreement takes the form of a pact through which the various parties commit to fulfil those responsibilities under the plan that fall within their remit and area of jurisdiction in order to achieve the strategic systems-level solution described within the STD programme.[10] An analogous consortium is applying the ideas developed by the network on multifunctional land use in the Winterswijk region on the Dutch–German border. Implementation is expected to cost 2 million guilders (US$1 million) over the first two years. The project is being partially funded privately and partially by the relevant municipal and provincial authorities. A third agreement concerns the redevelopment of a district of Rotterdam, where a covenant has been signed among relevant parties to restore and develop a river that is currently carried in an underground pipeline. The river will be elevated to a major role in the redesign of the district and will be restored to provide functions and attributes of benefit to local residents.

HAS THE PROGRAMME LEVERAGED ITS OWN EFFECTIVENESS?

There are several mechanisms by which the programme has sought to leverage its own effectiveness. One has been by directly exposing a very large number of professionals, opinion leaders and citizens of the Netherlands to the programme. The Programme Bureau employed around 30 individuals during its five-year term. Approximately 500 strategic planners, researchers, designers, engineers, architects or consultants were directly engaged in STD programme activities. In a still wider circle of influence, some 3,000–4,000 stakeholders took part in STD workshops and symposia. Members of these groups had repeated exposure to STD concerns and the STD innovation approach. The scale and range of the awareness and competence-building activities of the STD programme are seen both within and outside the programme as one of its major achievements. In an independent evaluation of the programme made for the German Parliament (Schramm and Wehling 1997), it is argued that the role of the STD programme in involving stakeholders in a societal

10   The major achievement is that all relevant agencies have committed to an action plan that sets out roles, responsibilities and a time plan. The advantage is not only that the pact provides a test bed for the technical solution developed within the STD programme, but also a test bed for the pact itself as an instrument for confidence building. It is particularly important to develop effective instruments since there are some 600 municipalities within the Netherlands that will be looking toward the Almelo experience in deciding the future of their own water systems.

debate about sustainable technology and in building awareness of the issues involved may, of itself, justify the programme.

Another mechanism has been that of developing and disseminating information about the programme and its approach, methodology and results during and after the programme, most recently through concentrated and targeted follow-up activities of executive training and sessions with industry, research institutes and government. The diffusion of the approach and method has partly been achieved through national and international contact networks of large or multinational companies, which was a deliberate intention of the STD programme design. Yet another mechanism has been building up a set of experienced individuals within the STD programme office, who have now left at the end of the programme to take positions in industry or to become consultants in sustainable technology development. Many of the 30 former staff of the Programme Bureau are now employed as consultants and are carrying forward STD-originated research and action plans or employing STD-developed methods in new contexts. There has also been a conscious attempt to forge international links and to stimulate joint working on innovation and sustainable technology research. In addition, the programme has been monitored in other countries with a view to the development of comparable national programmes.

Through these routes, the influence of the programme has spread beyond its immediate sphere of action to countries, companies and innovators who were not originally or directly involved in programme activities. International networks are now carrying forward several innovation fields pioneered in the STD programme. A project on sustainable households funded by the European Commission is continuing the lines of investigation set up in STD's laundry project. Joint working initiatives have been explored with China in relation to new energy technologies. Joint working has also been arranged with Japan, especially in the areas of deriving chemicals and materials from renewable feedstock, closed-cycle horticultural systems and sustainable cities. A joint Japanese–Dutch long-term research programme on using CO_2 as a renewable feedstock for industry has been proposed by NOW/SON.[11]

 FINAL REMARKS

Technology has an important role to play in the transition toward sustainable development. Technologies can contribute to sustainability by augmenting the resource base and by increasing resource productivity. But there is no guarantee

11   The initiative derives from the Dutch STD programme and the Japanese Earth Regeneration Plan.

that eco-efficient technologies will be developed in time, that they will be successful in replacing eco-inefficient technologies or that eco-efficiency gains will automatically be captured for environmental protection. Much depends on whether deliberate efforts are made to develop sustainable technologies, whether the necessary structural and cultural criteria for success are factored into innovation processes, whether innovation processes are inclusive, and how technologies are used. A strategic perspective is essential for identifying opportunities and for taking these up. Moreover, since sustainable technologies will only contribute to achieving sustainability if they are successful in replacing existing eco-inefficient solutions, attention needs to be paid early on to the social and structural factors that impinge on successful diffusion.

While it is clear from the STD case studies that sustainable technologies are possible, the programme makes equally clear that sustainable technologies are not easily developed in all fields of need and that substantial, conscious and consistent efforts are needed to search for sustainable technologies. Throughout the technology development process, there is need for a goal-oriented, strategic, co-evolutionary, systems perspective, which stresses the dynamic interrelation between cultural, structural and technological innovation. Technology is only one component of the development system. But it is a critical and potentially pivotal component. Proof that sustainable technologies are possible provides scope to make the needed structural economic changes for redirecting development. Given the long lead-times for innovation and diffusion of new technologies, there is no time to lose in beginning the search for sustainable technologies and systems solutions. The urgent need now is for innovation in regard to the innovation process itself.

Bibliography

A.D. Little Consultants (1993) *Final Report of the Definition Study on Novel Protein Foods* (Delft, Netherlands: STD).

Adriaanse, A., S. Bringezu, A. Hammond, Y. Moriguchi, E. Rodenberg, D. Rogich and H. Schütz (1997) *Resource Flows: The Material Basis of Industrial Economies* (Washington, DC: World Resources Institute).

Appleby, A. (1999) 'The Electrochemical Engine for Vehicles', *Scientific American*, July 1999: 58-63.

Argote, L., and D. Epple (1990) 'Learning Curves in Manufacturing', *Science* 247 (23 February 1990): 920-24.

Arthur, W. (1989) 'Competing Technologies, Increasing Returns, and Lock-in by Historical Events', *The Economic Journal* 99: 116-31.

Ayres, R. (1978) *Resources, Environment and Economics: Applications of the Materials/Energy Balance Principle* (New York: John Wiley).

Ayres, R. (1992) 'Toward a Non-Linear Dynamics of Technological Progress', *Journal of Economic Behavior and Organization* 24: 35-69.

Ayres, R. (1994) *Ecotransition in the Chemicals Industry* (CMER WP; Fontainebleau, France: INSEAD).

Ayres, R. (1998a) 'Ecorestructuring: The Transition to an Ecologically Sustainable Economy', in R. Ayres and P. Weaver (eds.), *Ecorestructuring: Implications for Sustainable Development* (Tokyo: United Nations University Press).

Ayres R. (1998b) *Turning Point: An End to the Growth Paradigm* (London: Earthscan).

Ayres, R., and A. Kneese (1971) 'Economic and Ecological Effects of a Stationary Economy', *Annual Review of Ecology and Systematics* 2 (April 1971).

Barra, L., and D. Coiante (1993) 'Energy Cost Analysis for Hydrogen-Photovoltaic Stand-Alone Power Stations', *International Journal of Hydrogen Energy* 18: 685-93.

Barsham, J. (1983) *Soil and Crop Science Society of Florida Proceedings* 42: 2-8.

Betz, F. (1993) *Strategic Technology Management* (New York: McGraw–Hill).

Borroni-Bird, C. (1996) 'Fuel Cell Commercialisation Issues for Light-Duty Vehicle Applications', *Journal of Power Sources* 61.1–2: 33-48.

Boulding, K. (1966) 'Environmental Quality in a Growing Economy', in J. Hardin (ed.), *Essays from the Sixth Resources for the Future (RFF) Forum* (Baltimore, MD: Johns Hopkins University Press).

Brooks, H. (1992) 'Sustainability and Technology', in N. Keyfitz (ed.), *Science and Sustainability* (Laxenburg, Austria: IIASA).

Büchel, K. (1990) 'The Role of the Chemical Industry for the Future of Mankind', paper for the *Chemistry and Industry Symposium of the Chemical Society of Japan*, January 1990.

Claesson, F. (1991) 'Causes and Scale of Dehydration', *Waterschapsbelangen* 91.23–24: 877-82.

CLTM (Dutch Committee on Long-Term Environmental Policy) (1990) *The Environment: Concepts for the 21st Century* (Zeist, Netherlands: Kerkebosch).

CLTM (Dutch Committee on Long-Term Environmental Policy) (1991) *Highlights from 'The Environment: Concepts for the 21st Century'* (Zeist, Netherlands: Kerkebosch).

Daly, H. (1990) 'Toward Some Operational Principles of Sustainable Development', *Ecological Economics* 2: 1-6.

De Haan, A., J. Quist and B. Linsen (1997) *Novel Protein Foods: Products with a Future* (Delft, Netherlands: STD).

DSM (1997) 'Sustainable Development and Fine Chemicals', *DSM Magazine* 72 (April 1997): 4-10.

Duijsens, L., and A. van der Schuyt (1996) 'New Technologies for Colouring Textiles', in *Sustainability and Chemistry* (Delft, Netherlands: STD).

Ehrlich, P., and A. Ehrlich (1990) *The Population Explosion* (London: Hutchinson).

Ehrlich, P., and J. Holdren (1971) 'Impact of Population Growth', *Science* 171: 1212-17.

Ehrlich, P., A. Ehrlich and J. Holdren (1977) *Ecoscience: Population, Resources, Environment* (San Francisco: W.H. Freeman).

Ekins, P., and M. Jacobs (1994) 'Environmental Sustainability and the Growth of GDP: Conditions for Compatibility', in V. Bhasskar and A. Glyn (eds.), *The North, the South and the Environment* (London: Earthscan).

Factor 10 Club (1994) *Carnoules Declaration* (Wuppertal, Germany: Wuppertal Institute for Climate, Energy and Environment).

Flowers, T., M. Hajibagheri and N. Clipson (1986) 'Halophytes', *The Quarterly Review of Biology* 61.3: 313-37.

Fortuin, J., and C. Boelhouwer (1996) 'New Reaction Media in Fine Chemistry in 2040', in *Sustainability and Chemistry* (Delft, Netherlands: STD).

Fourastié, J. (1949) *Le grand espoir du XXe siècle: Progrès technique, progrès économique, progrès social* (Paris: Presses Universitaires de France).

Frankel, M. (1955) 'Obsolescence and Technological Change in a Maturing Economy', *American Economic Review* 45: 269-319.

Freeman, C. (1994) 'The Economics of Technical Change', *Cambridge Journal of Economics* 18: 463-514.

Freeman, C., and C. Perez (1988) 'Structural Crisis of Adjustment: Business Cycles and Investment Behavior', in G. Dosi, C. Freeman, R. Nelson, G. Silverberg and L. Soete (eds.), *Technical Change and Economic Theory* (London/New York: Pinter).

Fussler, C., with P. James (1996) *Driving Eco-Innovation: A Breakthrough Discipline for Innovation and Sustainability* (London: Pitman).

GEA (Group for Efficient Appliances) (1995) *Basic Assumptions and Impact Analysis* (Denmark: GEA).

Gerben, Klein and Lebbink (eds.), (1994) *Microsystem Technology: Exploring Opportunities* (STT 56; Alphen aan den Rijn, Netherlands: Netherlands Study Centre for Technology Trends).

Gieling, T., and H. van den Vlekkert (1996) 'Application of ISFETs in Closed-Loop Systems for Greenhouses', *Advanced Space Research* 18.4–5: 135-38.

Goldemberg, J., T.B. Johansson, A.K.N. Reddy and R.H. Williams (1985) 'An End-Use Oriented Global Energy Strategy', *Annual Review of Energy* 10: 613-88.

Goodland, R., and H. Daly (1992) 'Ten reasons why northern income growth is not the solution to southern poverty', in R. Goodland, H. Daly and S. El Sarafi, *Population, Technology and Lifestyle: The Transition to Sustainability* (Washington, DC: Island Press).

Gouse, S., D. Gray, G. Tomlinson and D. Morrison (1992) 'Potential World Development through 2100: The Impacts on Energy Demand, Resources and the Environment', report to *The World Energy Council's 15th Congress*, Madrid, 1992.

Gray, P. (1989) 'The Paradox of Technological Development', in J. Ausubel *et al.* (eds.), *Technology and Environment* (Washington, DC: National Academy Press).

Gros, S. (1995) *Pyrolysis Oil as Diesel Fuel* (Vaasa, Finland: Wärtsilä Diesel International Ltd).

Grübler, A. (1992) *Technology and Global Change: Land-Use Past and Present* (IIASA WP-92-69; Laxenburg, Austria: IIASA).

Grübler, A. (1998) *Technology and Global Change* (Cambridge, UK: Cambridge University Press).

Hägerstrand, T. (1967) *Innovation Diffusion as a Spatial Process* (Chicago: University of Chicago Press).

Heij, G., and T. Schneider (1990) *Final Report of the 2nd Phase Dutch Priority Programme on Acidification* (Report no. 200-09; Bilthoven, Netherlands: RIVM).

Hetteling, J., R. Downing and P. de Smet (1991) *Mapping Critical Loads of Acidity for Europe* (Bilthoven, Netherlands: RIVM).

Hilhorst, M., and C. Dirksen (1994) 'Di-electric Water Content Sensors: Time Domain versus Frequency Domain', *Proceedings of the Symposium at Northwestern University*, Bureau of Mines Special Publication SP19-94, Washington, DC.

Holdren, J., and P. Ehrlich (1974) 'Human Population and the Global Environment', *American Scientist* 62: 282-92.

Holdren, H., and R. Pachauri (1991) 'Energy', paper presented at the *Ascend Conference*, ICSU, Vienna, November 1991.

Hueting, R., P. Bosch and B. de Boer (1992) *Methodology for the Calculation of Sustainable National Income* (M-44; The Hague: Central Statistical Bureau).

IIASA (International Institute for Applied Systems Analysis) and WEC (World Energy Council) (1995) *Global Energy Perspectives to 2050 and Beyond* (London: WEC).

IUCN (1983) *List of Rare, Threatened and Endemic Plants in Europe* (IUCN Threatened Plant Unit; Strasbourg: Council of Europe).

IPCC (Intergovernmental Panel on Climate Change) (1994) *Radiative Forcing of Climate Change: The 1994 Special Report of the Scientific Assessment Working Group of the IPCC* (ed. J. Houghton, L. Meiro-Filho, J. Bruce, B. Hoesung-Lee, B. Callander, E. Haites, N. Harris and K. Maskell; Cambridge, UK: Cambridge University Press).

IPCC (Intergovernmental Panel on Climate Change) (1996a) *Climate Change 1995: The Science of Climate Change* (Contribution of Working Group I to the Second Assessment Report of the IPCC; ed. J. Houghton, L. Meiro-Filho, B. Callander, N. Harris, A. Kattenberg and K. Maskell; Cambridge, UK/New York: Cambridge University Press).

IPCC (Intergovernmental Panel on Climate Change) (1996b) *Climate Change 1995: Impacts, Adaptations and Mitigation of Climate Change. Scientific-Technical Analysis* (Contribution of Working Group II to the Second Assessment Report of the IPCC; ed. R. Watson, M. Zinyowera and R. Moss; Cambridge, UK/New York: Cambridge University Press).

Jansen, L. (1994) 'Towards a Sustainable Future: En Route with Technology', in CLTM (ed.), *The Environment: Towards a Sustainable Future* (Dordrecht: Kluwer).

Jansen, L. (1996a) 'A Challenge for Government: How to Dematerialise?', paper presented at the *Stockholm Seminar on Strategy towards Sustainable Development*, June 1996.

Jansen, L. (1996b) 'Technology for Sustainable Business in the 21st Century', paper presented at the *Jakarta Symposium on Environmental Audits*, August 1996.

Jansen, L., and J. Don (1996) *Chemistry in Sustainable Development* (STD WP; Delft, Netherlands: STD).

Janssons, R., L. Sijtsma and A. Linnemann (1996) *Voedingswaarde van Novel Protein Foods* (STD WP-VN-14; Delft, Netherlands: STD).

Johansson, T., H. Kelly, A. Reddy and R. Williams (1993) *Renewable Energy Sources for Fuels and Electricity* (Washington, DC: Island Press).

Judson, S. (1986) 'Erosion of our Land or What's happening to our continent?', *American Scientist* 56: 356-74.

Kampers, F., J. Bakker, G. Bot, C. Eerkens and O. de Kuijer (1995) *Sensors for Agricultural Production in the 21st Century* (report for the STD Programme; Wageningen: Institute of Agricultural and Environmental Engineering).

Kline, S. (1985) 'What is technology?', *Bulletin of Science, Technology and Society* 5.3: 215-19.

Kondratiev, N. (1926) 'Die Langenwellen in der Konjunktur', *Arkiv für Sozialwissenschaft und Sozialpolitik* 56: 573-609.

Kouffeld, R. (1990) 'World Energy Supplies and the Consequences for the Environment', paper presented at the Technical University of Delft (TNO).

Krause, F., W. Bach and J. Kooney (1990) *Energy Policy in the Greenhouse: From Warning Fate to Warning Limit* (London: Earthscan).

Kruijsen, J., R. Meijkemp and A. Zweers (1994) *Service Products and Sustainable Development: Report of the Workshop at Delft University of Technology* (Delft, Netherlands: DUT Faculty of Industrial Design Engineering).

Lee, T., H. Linden, D. Dreyfus and T. Vasko (1988) *The Methane Age* (GeoJournal Library; Dordrecht: Kluwer).

Marchetti, C., and N. Nakicenovic (1979) *The Dynamics of Energy Systems and the Logistic Substitution Model* (IIASA RR-79-13; Laxenburg, Austria: IIASA).

Meijer, H., and J. van Leeuwen (1997) *The Sustainable Urban Water Cycle: End Report* (STD WP-W4; Delft, Netherlands: STD).

Morone, J. (1993) *Winning in High-Tech Markets* (Boston, MA: Harvard Business School Press).

Morris, D., and I. Ahmed (1992) *The Carbohydrate Economy* (Washington, DC: Institute for Local Self-Reliance).

Moser, A. (1998) 'Ecological Process Engineering: The Potential of Bioprocessing', in R. Ayres and P. Weaver (eds.), *Ecorestructuring: Implications for Sustainable Development* (Tokyo: United Nations University Press).

Mulder, K. (1992) *Choosing the Corporate Future: Technology Networks and Choice Concerning the Creation of High-Performance Fibre Technology* (Groningen, Netherlands: University of Groningen).

Mulder, K., C. van de Weijer and V. Marchau (1996) *Expert Opinions on the Future of Vehicle Propulsion: Report of a Delphi Study* (STD-WP-M3; Delft, Netherlands: STD).

Mulderink, J. (1997) 'A Sustainable Future via Chemical Technology', paper presented at the *Achema Symposium*, Frankfurt, June 1997.

Nakicenovic, N., A. Grübler, S. Inaba, S. Messner, S. Nilson, Y. Nishimura, H.-H. Rogner, A. Schäfer, L. Schrattenholzer, M. Strubegger, J. Swisher, D. Victor and D. Wilson (1993) 'Long-Term Strategies for Mitigating Global Warming', *Energy* 18.5 (Special Issue).

Nelson, R., and S. Winter (1977) 'In Search of a Useful Theory of Innovations', *Research Policy* 6: 36-77.

Nelson, R., and S. Winter (1982) *An Evolutionary Theory of Economic Change* (Cambridge, MA: Harvard University Press).

NEPP (Netherlands National Environmental Policy Plan) (1989) *To Choose or to Lose* (Bilthoven, Netherlands: RIVM).

Nilsson, J., and P. Grennfeldt (1988) *Critical Loads for Sulphur and Nitrogen* (Stockholm: Nordic Council of Ministers).

Okkerse, C., and H. van Bekkum (1996) 'Renewable Raw Materials for the Chemicals Industry', in *Sustainability and Chemistry* (Delft, Netherlands: STD).

Opschoor, J., and L. Reijnders (1991) 'Toward Sustainable Development Indicators', in O. Kuik and H. Verbruggen (eds.), *In Search of Indicators of Sustainable Development* (Dordrecht: Kluwer): 7-28.

Opschoor, J., and F. van der Ploeg (1991) 'Sustainability and Quality: Main Objectives of Environmental Policy', in CLTM (ed.), *The Environment: Concepts for the 21st Century* (Zeist, Netherlands: Kerkebosch).

Pezzey, J, (1989) *Sustainability, Intergenerational Equity and Environmental Policy* (Discussion Paper in Economics 89-7; Boulder, CO: University of Colorado, Department of Economics).

Prins, W., and W. van Swaaij (1998) *Flash Pyrolysis Oil: A Market Study* (Report no. 1494; Utrecht: Novem).

Prins, W., and W. van Swaaij (1999) 'Biomass Flash Pyrolysis', in *Twente Presentation*, Japan, February 1999.

Quakernaat, J. (1995) 'Hydrogen in a Long-Term Perspective', *International Journal of Hydrogen Energy* 20.6: 485-92.

Quakernaat, J., and L. Vermij (1994) 'Energy Master: Planning for Energification Purposes', in T.N. Veziroglu (ed.), *Renewable Energy Sources: International Progress*. Part B (Amsterdam: Elsevier Science).

Quist, J., O. de Kuijer, A. de Haan, H. Linsen, H. Hermans and I. Larsen (1996) 'Restructuring Meat Consumption: Novel Protein Foods in 2035', paper presented at the *Greening of Industry Conference*, Heidelberg, 24–27 November 1996.

Reddy, A., and J. Goldemberg (1990) 'Energy for the Developing World', *Scientific American*, September 1990 (Special Issue).

Repetto, R., W. Magrath, M. Wells, C. Beer and F. Rossini (1989) *Wasting Assets: Natural Resources in the National Income Accounts* (Washington, DC: World Resources Institute).

Rip, A. (1989) 'Expectations and Strategic Niche Management in Technological Development', paper presented at *Inside Technology Conference*, Turin, June 1989.

Rip, A., T. Misa and J. Schot (eds.) (1995) *Managing Technology in Society: The Approach of Constructive Technology Assessment* (London: Pinter).

RIVM (Netherlands Ministry of Housing, Physical Planning and the Environment) (1988) *Concern for Tomorrow: The First National Environmental Outlook* (Bilthoven, Netherlands: RIVM).

RIVM (Netherlands Ministry of Housing, Physical Planning and the Environment) (1991) *National Environmental Survey 1990-2010* (Bilthoven, Netherlands: RIVM).

RIZA (Institute for Inland Water Management and Wastewater Treatment) (1996/1997) *Report to STD on Results from Runs of the PROMISE Model* (Bilthoven, Netherlands: RIVM).

Robbelen, G., *et al.* (ed.) (1991) *Oil Crops in the World* (New York: McGraw–Hill).

Rogner, H.-H. (1990) 'Analyse der Förderpotentiale und langfristigen Verfürbarkeit von Kohle, Erdgas und Erdöl', in *A Report of the Committee of Inquiry of the German Parliament into Protection of the Atmosphere: Energy and Climate: Fossil Energy Carriers* (Bonn: Economica Verlag).

Rogner, H.-H. (1997) 'An Assessment of the World Hydrocarbon Resources', *Annual Review of Energy and the Environment* 22: 217-62.

Rogner, H.-H. (1998) 'Global Energy Futures: The Long-Term Perspective for Ecorestructuring', in R. Ayres and P. Weaver (eds.), *Ecorestructuring: Implications for Sustainable Development* (Tokyo: United Nations University Press).

Rohatgi, P., K. Rohatgi and R. Ayres (1998) 'Materials Future: Pollution Prevention, Recycling and Improved Functionality', in R. Ayres and P. Weaver (eds.), *Ecorestructuring: Implications for Sustainable Development* (Tokyo: United Nations University Press).

Rosegger, G. (1996) *The Economics of Production and Innovation: An Industrial Perspective* (Oxford, UK: Butterworth–Heinemann, 3rd edn).

Rosenburg, N. (1982) *Inside the Black Box: Technology and Economics* (Cambridge, UK: Cambridge University Press).

Rozema, J., H.-J. Wiesenekker, R. Decae and J. Broerse (1996) 'Agricultural Production on Brackish Soil in 2040', in *Sustainability and Chemistry* (Delft, Netherlands: STD).

Schot, J. (1992) 'Constructive Technology Assessment and Technology Dynamics: The Case of Clean Technologies', *Science, Technology and Human Values* 17.1: 36-56.

Schot, J., and A. Rip (1996) 'The Past and Future of Constructive Technology Assessment', *Technological Forecasting and Social Change* 54: 251-68.

Schramm, E., and P. Wehling (1997) *Forschungspolitik für eine nachhaltige Entwicklung: Das niederländische DTO-Programm und seine Bedeutung für die Bundesrepublik Deutschland* (Frankfurt: Institut für sozial-ökologische Forschung).

Schumpeter, J. (1911) *Theorie der wirtschaftlichen Entwicklung* (Leipzig, Germany: Duncker & Humblot).

Schumpeter, J. (1934) *The Theory of Economic Development: An Inquiry into Profits, Capital, Credit, Interest and the Business Cycle* (Cambridge, MA: Harvard University Press).

Schumpeter, J. (1935) 'The Analysis of Economic Change', *Review of Economic Statistics* 17: 2-10.

Schumpeter, J. (1939) *Business Cycles: A Theoretical, Historical and Statistical Analysis of the Capitalist Process* (2 vols.; New York: McGraw–Hill).

Schumpeter, J. (1942) *Capitalism, Socialism and Democracy* (New York: Harper & Bros).

Schwarz, M. (1992) *Report to the Programme Director, Environment and Technology Ministry* (The Hague: VROM-DGM, February 1992).

Schwarz, M., and M. Thomson (1990) *Divided We Stand* (New York: Harvester Wheatsheaf).

Simonis, U. (1994) 'Industrial Restructuring in Industrial Countries', in R. Ayres and U. Simonis (eds.), *Industrial Metabolism: Restructuring for Sustainable Development* (Tokyo: United Nations University Press).

Smith, M., and L. Marx (1994) *Does technology drive history?* (Cambridge, MA: MIT Press).

Smits, R., and J. Leyten (1991) *Technology Assessment* (Zeist, Netherlands: Kerkebosch).

Snel, J., O. van Kooten and L. van Hove (1991) 'Assessment of Stress in Plants by Analysis of Photosynthetic Performance', *Transactions of Analytical Chemistry* 10.1: 26-30.

Solow, R. (1986) 'On the Intergenerational Allocation of Natural Resources', *Scandinavian Journal of Economics* 88: 141-49.

TERES (The European Renewable Energy Study) (1994) *Technology Profiles*. Vol. 2 Annex 1 (Brussels: Commission of the European Communities).

Tietenberg, T. (1984) *Environmental and Natural Resource Economics* (Glenview, IL: Scott Foresman).

TNO-MEP (1997) *Milieu-Analyse van Verbeteropties* (STD WP-W3; Delft, Netherlands: STD).

Tweede Kamer (1991) *Beleidsnotie. Technologie en Milieu: 1990–1991* (Report 22085, 1991, no 1; The Hague).

USOTA (United States Office of Technology Assessment) (1992) 'Agricultural Commodities as Industrial Raw Materials', *Chemical Marketing Reporter*, 9 March 1992.

Van den Akker, C., P. Terpstra, J. Wiggers and G. Lettinga (1997) *Essays on the Sustainable Urban Water Cycle* (STD WP-W2; Delft, Netherlands: STD).

Van den Berg, N., G. Huppers, B. van der Ven and B. Krutwagen (1996) *Novel Protein Foods: An Environmental Analysis* (STD WP-VN-18; Delft, Netherlands: STD).

Van den Hoed, R., and P. Vergragt (1996) *Toward Sustainable Laundering of Clothes: Report of the Workshop of 27 September 1996* (Delft, Netherlands: STD).

Van Eerdt, Oltshoorn and Smeets (1993) *Energieverbruik, ruimtebeslag en mineralen in de Europese varkenshouderij.*

Van Leeuwen, J., and H. Meijer (1996) *Illustration Process of a Sustainable Urban Water Cycle* (STD-WP-W1; Delft, Netherlands: STD).

Veblen, T. (1904) *The Theory of Business Enterprise* (New York: Scribner).

Veblen, T. (1921) *The Engineers and the Price System* (New York: Huebsch).

Veblen, T. (1953) *The Theory of the Leisure Class: An Economic Study of Institutions* (New York: New American Library, rev. edn).

Vergragt, P. (1988) 'The Social Shaping of Industrial Innovations', *Social Studies of Science* 18: 483-513.

Vergragt, P., and G. Grootveld (1994) 'Sustainable Technology Development in the Netherlands: The First Phase of the Dutch STD Programme', *Journal of Cleaner Production* 2.3-4: 133-37.

Vergragt, P., and L. Jansen (1993) 'Sustainable Technological Development: The Making of a Dutch Long-Term-Oriented Technology Programme', *Project Appraisal* 8.3: 134-40.

Vergragt, P., and M. van der Wel (1998) 'The Backcasting Approach: Sustainable Washing as an Example', in N. Roome (ed.), *Sustainable Strategies for Industry* (Washington, DC: Island Press).

Vergragt, P., and D. van Noort (1996) 'Sustainable Technology Development: The Mobile Hydrogen Fuel Cell', *Business Strategy and the Environment* 5: 168-77.

Vergragt, P., M. van der Wel, R. Meykamp and M. Tassoul (1995) 'Involving Industry in Backcasting Scenario Building', paper presented at the *4th Greening of Industry Conference*, Toronto, 12–14 November 1995.

Vermeulen, J. (1996) 'Light Materials in Strong Structures', in *Sustainability and Chemistry* (Delft, Netherlands: STD).

Wagenaar, B., W. Prins and W. van Swaaij (1993) 'Flash Pyrolysis Kinetics of Pine Wood', *Fuel Processing Technology* 36: 291-98.

Wagenaar, B. (1994) PhD thesis, University of Twente, Netherlands.

WCED (World Commission on Environment and Development) (1987) *Our Common Future* (Oxford, UK: Oxford University Press).

Weaver, P.M. (1995) 'Steering the Eco-transition: A Material Accounts Approach', in M. Taylor (ed.), *Environmental Change: Industry, Power and Policy* (Aldershot, UK: Avebury).

Weaver, P.M. (1997) 'Why businesses and nations would benefit from leadership in resource productivity improvement', paper for the *International Factor 10 Club Annual Meeting*, Carnoules, France.

Weissermel, K., and H.-J. Arpe (1993) *Industrial Organic Chemistry* (Weinheim: VCH, 2nd edn).

Westing, E. (1981) 'A World in Balance', *Environmental Conservation* 8.3: 177-83.

Weterings, R., and J. Opschoor (1992) *The Ecocapacity as a Challenge to Technology Development* (Advisory Council on Nature and the Environment [RMNO] Report no. 74A; Rijswijk, Netherlands: RMNO).

Weterings, R., and J. Opschoor (1994) *Toward Environmental Performance Indicators Based on the Notion of Environmental Space* (Advisory Council on Nature and the Environment [RMNO] Report no. 96; Rijswijk, Netherlands: RMNO).

Williams, R. (1998) 'Fuel Decarbonisation for Fuel-Cell Applications and Sequestration of the Separated Carbon Dioxide', in R. Ayres and P. Weaver (eds.), *Ecorestructuring: Implications for Sustainable Development* (Tokyo: United Nations University Press).

Williams, R., E. Larson, R. Katofsky and J. Chen (1995) 'Methanol and Hydrogen from Biomass for Transportation', *Energy for Sustainable Development* 1.5: 18-34.

Wilson, E. (1989) 'Threats to Biodiversity', *Scientific American*, September 1989: 60-66.

Winter, C., and M. Fuchs (1991) 'Hysolar and Solar-Wasserstoff-Bayern', *International Journal of Hydrogen Energy* 16: 723-34.

Witteveen and Bos Consultants (1995) *The Sustainable Urban Water Cycle: An Exploratory Study* (Delft, Netherlands: STD).

Wright, D. (1990) 'Human Impacts on Energy Flows through Natural Ecosystems and Implications for Species Endangerment', *Ambio* 19.4: 189-94.

Wurster, R., and W. Zittel (1994) *Energy Technologies to Reduce Carbon Dioxide Emissions in Europe: Prospects, Competition, Synergy* (Paris: IEA/OECD).